A2-Level
Physics

A2 Physics is seriously tricky — no question about that.
To do well, you're going to need to revise properly and practise hard.

This book has thorough notes on all the theory you need,
and it's got practice questions... lots of them.
For every topic there are warm-up and exam-style questions.

And of course, we've done our best to make the whole thing vaguely entertaining for you.

Complete Revision and Practice
Exam Board: OCR B

Editors:
Amy Boutal, Sarah Hilton, Alan Rix, Julie Wakeling, Sarah Williams

Contributors
Stuart Barker, Jane Cartwright, Peter Cecil, Mark A. Edwards, Barbara Mascetti, John Myers, Zoe Nye, Moira Steven, Andy Williams, Tony Winzor

Proofreaders:
Ian Francis, Glenn Rogers

Published by CGP

ISBN: 978 1 84762 272 3

Groovy website: www.cgpbooks.co.uk
Jolly bits of clipart from CorelDRAW®
Printed by Elanders Ltd, Newcastle upon Tyne.

Based on the Classic CGP style created by Richard Parsons.

Contents

The Scientific Process

'How Science Works' is all about the scientific process — how we develop and test scientific ideas.
It's what scientists do all day, every day (well, except at coffee time — never come between a scientist and their coffee).

Scientists Come Up with **Theories** — Then **Test Them**...

Science tries to explain **how** and **why** things happen — it **answers questions**. It's all about seeking and gaining **knowledge** about the world around us. Scientists do this by **asking** questions and **suggesting** answers and then **testing** them, to see if they're correct — this is the **scientific process**.

1.) **Ask** a question — make an **observation** and ask **why or how** it happens.
E.g. what is the nature of light?

2) **Suggest** an answer, or part of an answer, by forming:

- a **theory** (a possible **explanation** of the observations)
e.g. light is a wave.
- a **model** (a **simplified picture** of what's physically going on)

3) Make a **prediction** or **hypothesis** — a **specific testable statement**, based on the theory, about what will happen in a test situation.
E.g. light should interfere and diffract.

4) Carry out a **test** — to provide **evidence** that will support the prediction, or help disprove it. E.g. Young's double-slit experiment.

The evidence supported Quentin's Theory of Flammable Burps.

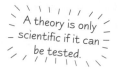

A theory is only scientific if it can be tested.

...Then They **Tell** Everyone About Their **Results**...

The results are **published** — scientists need to let others know about their work. Scientists publish their results in **scientific journals**. These are just like normal magazines, only they contain **scientific reports** (called papers) instead of the latest celebrity gossip.

1) Scientific reports are similar to the **lab write-ups** you do in school. And just as a lab write-up is **reviewed** (marked) by your teacher, reports in scientific journals undergo **peer review** before they're published.

2) The report is sent out to **peers** — other scientists that are experts in the **same area**. They examine the data and results, and if they think that the conclusion is reasonable it's **published**. This makes sure that work published in scientific journals is of a **good standard**.

3) But peer review **can't guarantee** the science is **correct** — other scientists still need to **reproduce** it.

4) Sometimes **mistakes** are made and bad work is published. Peer review **isn't perfect** but it's probably the best way for scientists to self-regulate their work and to publish **quality reports**.

...Then **Other Scientists** Will **Test** the Theory Too

Other scientists read the published theories and results, and try to **test the theory** themselves. This involves:

- Repeating the **exact same experiments**.
- Using the theory to make **new predictions** and then testing them with **new experiments**.

If the **Evidence** Supports a Theory, It's **Accepted** — for Now

1) If all the experiments in all the world provide evidence to back it up, the theory is thought of as **scientific 'fact'** (for now).

2) But they never become **totally undisputable** fact. Scientific **breakthroughs or advances** could provide new ways to question and test the theory, which could lead to **new evidence** that **conflicts** with the current evidence. Then the testing starts all over again...

And this, my friend, is the **tentative nature of scientific knowledge** — it's always **changing** and **evolving**.

The Scientific Process

So scientists need evidence to back up their theories. They get it by carrying out experiments, and when that's not possible they carry out studies. But why bother with science at all? We want to know as much as possible so we can use it to try and improve our lives (and because we're nosey).

Evidence Comes From Controlled Lab Experiments...

1) Results from **controlled experiments** in **laboratories** are **great**.
2) A lab is the easiest place to **control variables** so that they're all **kept constant** (except for the one you're investigating).

> For example, finding the charge stored on a capacitor by charging at a constant current and measuring the voltage across it (see p. 6). All other variables need to be kept the same, e.g. the current you use and the temperature, as they may also affect its capacitance.

... That You can Draw Meaningful Conclusions From

"Right Geoff, you can start the experiment now... I've stopped time..."

1) You always need to make your experiments as **controlled** as possible so you can be confident that any effects you see are linked to the variable you're changing.
2) If you do find a relationship, you need to be careful what you conclude. You need to decide whether the effect you're seeing is **caused** by changing a variable, or whether the two are just **correlated**.

Society Makes Decisions Based on Scientific Evidence

1) Lots of scientific work eventually leads to **important discoveries** or breakthroughs that could **benefit humankind**.
2) These results are **used by society** (that's you, me and everyone else) to **make decisions** — about the way we live, what we eat, what we drive, etc.
3) All sections of society use scientific evidence to make decisions, e.g. politicians use it to devise policies and individuals use science to make decisions about their own lives.

Other factors can **influence** decisions about science or the way science is used:

Economic factors
- Society has to consider the **cost** of implementing changes based on scientific conclusions — e.g. the cost of reducing the UK's carbon emissions to limit the human contribution to **global warming**.
- Scientific research is often **expensive**. E.g. in areas such as astronomy, the Government has to **justify** spending money on a new telescope rather than pumping money into, say, the **NHS** or **schools**.

Social factors
- **Decisions** affect **people's lives** — e.g. when looking for a site to build a **nuclear power station**, you need to consider how it would affect the lives of the people in the **surrounding area**.

Environmental factors
- Many scientists suggest that building **wind farms** would be a **cheap** and **environmentally friendly** way to generate electricity in the future. But some people think that because **wind turbines** can **harm wildlife** such as birds and bats, other methods of generating electricity should be used.

So there you have it — how science works...

Hopefully these pages have given you a nice intro to how science works, e.g. what scientists do to provide you with 'facts'. You need to understand this, as you're expected to know how science works yourself — for the exam and for life.

Radioactivity and Exponential Decay

Making models in physics is nothing to do with squeezy bottles and sticky-back plastic, I'm afraid. But they are designed to make life easier — they simplify things and show links between topics you might not think are connected.

Models Use Assumptions to Simplify Problems

1) A **model** is a **set of assumptions** that **simplifies** and **idealises** a particular problem.

2) The assumptions mean you can write **equations** that describe the processes, and so make **calculations** and **predictions**. Without the assumptions, there would be **too many factors** to consider.

3) The topics in this section may appear to be **unconnected**, but they're all based on **models** that use **differential equations** to describe the **rate of change** of something.

4) This is another **benefit** of models — once you have a model that describes one process you can often **extend it**, or **change it slightly**, to describe **another**, **unrelated process** without having to start all over again — compare the models for **radioactive decay** and **capacitor discharge** and you'll see what I mean.

Unstable Atoms are Radioactive

1) If an atom has **too many neutrons**, **not enough neutrons**, or **too much energy** in the nucleus, it may be **unstable**.

2) Unstable atoms **break down** by **releasing energy** and/or **particles**, until they reach a **stable form** — this process is called **radioactive decay**.

3) Radioactive decay is a **random** process — you **can't** tell when any **one atom** will **decay**, or **which atom** in a sample will be the **next** to decay.

It could be you.

Radioactivity can be Modelled by Exponential Decay

1) So, if you can't predict when an atom will decay, how do you do any **calculations** or make any **predictions**? The answer is to use a **model** — exponential decay — based on a **very large number** of atoms.

2) If you take a **large** enough **sample** of **unstable atoms**, the **overall behaviour** shows a **pattern**. You can't predict the decay of an **individual atom**, but you can predict **how many atoms** will decay in a **given time**.

3) If you plot a graph showing the **number** of atoms that **decay each second** against **time**, you always get an **exponential decay curve** — like the one on page 5 — so this can be used as a **model** for radioactive decay.

The Rate of Decay is Measured by the Decay Constant

The **number** of unstable atoms that **decay each second** is called the **activity** of the sample. The **activity** of a sample is **proportional** to the **size of the sample** — which is why an **activity-time graph** is an **exponential decay** curve. As **atoms decay**, the **sample** size gets **smaller**, so the **activity falls** and the **graph** gets **shallower and shallower**.

The **decay constant** (λ) measures how **quickly** an isotope will **decay** — it's the **probability** of a given nucleus **decaying** in a certain **time**. The **bigger** the value of λ, the **more likely** a decay is, so the **faster** the rate of decay. Its unit is s^{-1}.

| activity = decay constant × number of atoms |

Or in symbols: $A = \lambda N$

Don't get λ confused with wavelength.

Activity is measured in **becquerels** (Bq): | 1 Bq = 1 decay per second (s^{-1}) |

N is the number of radioactive nuclei remaining in the sample.

If you graph the number of unstable atoms remaining in a sample (N) against time (see next page), the **gradient** of the graph is **negative**. The gradient is the change in the number of radioactive nuclei remaining in a given time (the rate of decay), which must also be **negative**.

$$\frac{dN}{dt} = -\lambda N$$

Equations like $\frac{dN}{dt} = -\lambda N$ are called **differential equations**. They're used to describe the **rate of change** of something — in this case it's the **number of undecayed atoms** in a sample, but this model can be used in **many other** situations. For example, the **decay of charge** on a **capacitor** (pages 6-7) is described by a **very similar** equation.

Radioactivity and Exponential Decay

You can Find the Half-Life of an Isotope from the Graph

The **half-life** ($T_{1/2}$) of an **isotope** is the **average time** it takes for the **number of undecayed atoms** to **halve**.

In **practice**, half-life isn't measured by counting atoms (which can be very tricky), but by measuring the **time it takes** for the **activity** to **halve**. The **longer** the **half-life** of an isotope, the **longer** it stays **radioactive**.

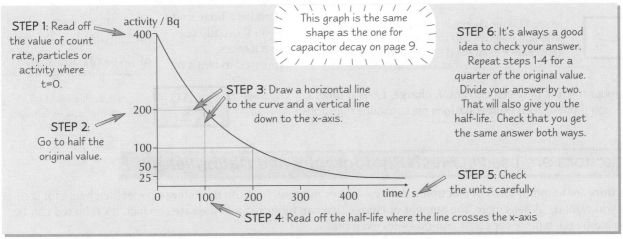

STEP 1: Read off the value of count rate, particles or activity where t=0.

STEP 2: Go to half the original value.

STEP 3: Draw a horizontal line to the curve and a vertical line down to the x-axis.

This graph is the same shape as the one for capacitor decay on page 9.

STEP 6: It's always a good idea to check your answer. Repeat steps 1-4 for a quarter of the original value. Divide your answer by two. That will also give you the half-life. Check that you get the same answer both ways.

STEP 5: Check the units carefully

STEP 4: Read off the half-life where the line crosses the x-axis

When you're **measuring** the **activity** and **half-life** of a **source**, you've got to **remember background radiation**. The **background radiation** needs to be **subtracted** from the **activity** readings to give the **source activity**.

You Need to Know the Equations for Half-Life and Decay...

1) The **half-life** can be **calculated** using the equation:
(where ln is the natural log)

$$T_{\frac{1}{2}} = \frac{\ln 2}{\lambda}$$

2) The **number of radioactive atoms** remaining, **N**, depends on the **number originally** present, **N_o**. The **number remaining** can be calculated using the equation:

$$N = N_0 e^{-\lambda t}$$

Here t = time, measured in seconds.

This is the **exact solution** of the differential equation $dN/dt = -\lambda N$.

Example:
A sample of the radioactive isotope ^{13}N contains 5×10^6 atoms. The decay constant for this isotope is 1.16×10^{-3} s^{-1}.

a)   What is the half-life for this isotope?

$$T_{\frac{1}{2}} = \frac{\ln 2}{1.16 \times 10^{-3}} = 598 \text{ s}$$

b)   How many atoms of ^{13}N will remain after 800 seconds?

$$N = N_0 e^{-\lambda t} = 5 \times 10^6 \, e^{-(1.16 \times 10^{-3})(800)} = 1.98 \times 10^6 \text{ atoms}$$

Practice Questions

Q1   Define radioactive activity. What units is it measured in?

Q2   Sketch a general radioactive decay graph showing the number of undecayed particles against time.

Q3   What is meant by the term 'half-life'?

Exam Question

Q1   A pure radioactive source initially contains 50 000 atoms.
The decay constant for the sample is $\lambda = 0.014$ ms^{-1}.

(a) What is the half-life of this sample?   [2 marks]

(b) Approximately how many atoms of the radioactive source will there be after 300 seconds?   [2 marks]

(c) Exponential decay is used to model radioactive decay. Give one reason why models are useful in science.   [1 mark]

Radioactivity is a random process — just like revision shouldn't be...

Remember the shape of that graph — whether it's count rate, activity or number of atoms plotted against time, the shape's always the same. This is all pretty straightforward mathsy-type stuff: plugging values in equations, reading off graphs, etc. Not very interesting, though. Ah well, once you get onto classifying particles you'll be longing for a bit of boredom.

Capacitors

You might not have guessed that a page on radioactive decay would be followed by a couple of pages on capacitors, but by the time you get to pages 10 and 11 it'll make perfect sense. Anyway, first you need to learn about capacitors...

Capacitance is Defined as the Amount of Charge Stored per Volt

Capacitors are things that **store electrical charge** — a bit like a charge bucket. **Capacitance** is a measure of **how much charge** a capacitor can hold — it's defined as the **amount of charge stored per volt**.

$$C = \frac{Q}{V}$$

where Q is the **charge** in coulombs, V is the **potential difference** in volts and C is the **capacitance** in farads (F) — 1 farad = 1 C V^{-1}.

A farad is a **huge** unit so you'll usually see capacitances expressed in terms of:

μF — microfarads ($\times 10^{-6}$)
nF — nanofarads ($\times 10^{-9}$)
pF — picofarads ($\times 10^{-12}$)

Remember the link between **current**, I, **charge**, Q, and **time**, t, too — you'll often need **both equations** for capacitor questions.

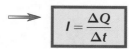
$$I = \frac{\Delta Q}{\Delta t}$$

You met this equation at AS, so it should be familiar.

Capacitors are used in Flash Photography and Defibrillators

Capacitors are found in loads of **electronic devices** — they're useful because they **store** up electric charge for use when you want it. What's more, the **amount of charge** that can be stored and the **rate** at which it's **released** can be controlled by the type of capacitor chosen — for example:

1) **Flash photography** — when you take a picture, charge stored in a **capacitor** flows through a tube of xenon gas which emits a **bright light**. In order to give a **brief flash** of bright light, the capacitor has to discharge really quickly to give a **short pulse** of high current.

2) **Defibrillators** — during a **heart attack**, a person's heart **stops beating** properly. A **proper rhythm** can be restarted by giving the person an **electric shock** using a **defibrillator**. The circuit in a defibrillator can be programmed to **vary** how much **charge** is stored depending on the size of the patient. The charge is stored on a **capacitor** and then released in a **short, controlled burst**.

3) **Back-up power supplies** — **computers** are often connected to back-up power supplies to make sure that you don't lose any data if there's a **power cut**. These often use **large capacitors** that store charge while the power is on then release that charge slowly if the power goes off. The capacitors are designed to discharge over a number of hours, maintaining a **steady flow** of charge.

You can Investigate the Charge Stored by a Capacitor Experimentally

Investigating the Charge Stored on a Capacitor

1) Set up a **test circuit** to measure current and potential difference:

2) Constantly adjust the **variable resistor** to keep the charging current **constant** for as long as you can (it's impossible when the capacitor is nearly fully charged).

3) Record the p.d. at regular intervals until it **equals** the **battery p.d.**

4) From these results, you can plot the following graphs:

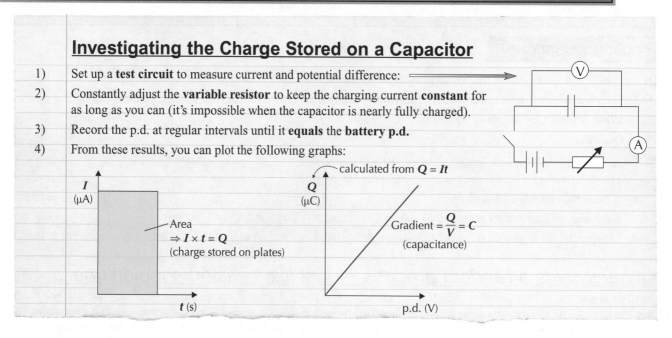

Area $\Rightarrow I \times t = Q$ (charge stored on plates)

calculated from $Q = It$

Gradient $= \dfrac{Q}{V} = C$ (capacitance)

Capacitors

Capacitors **Store Energy**

1) In this circuit, when the switch is flicked to the **left**, **charge** builds up on the plates of the **capacitor**. **Electrical energy**, provided by the battery, is **stored** by the capacitor.

2) If the switch is flicked to the **right**, the energy stored on the plates will **discharge** through the **bulb**, converting electrical energy into light and heat.

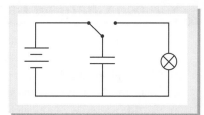

3) **Work** is done **removing charge** from **one plate** and depositing **charge** onto the other one. The energy for this must come from the **electrical energy** of the **battery**, and is given by **charge × p.d.**

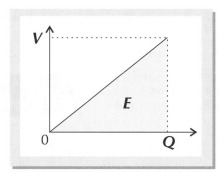

4) You can find the **energy stored** by the capacitor from the **area** under a **graph** of **p.d.** against **charge stored** on the capacitor.

The p.d. across the capacitor is **proportional** to the charge stored on it (see p. 6), so the graph is a **straight line** through the origin.

The **energy stored** is given by the **yellow triangle**.

5) **Area of triangle = ½ × base × height**, so the energy stored by the capacitor is:

$$E = \frac{1}{2}QV$$

You Need to Know **Two** Expressions for the **Energy Stored** by a Capacitor

1) You know the first one already: $E = \dfrac{1}{2}QV$

2) $C = \dfrac{Q}{V}$, so $Q = CV$. Substitute that into the energy equation: $E = \dfrac{1}{2}CV \times V$. So: $E = \dfrac{1}{2}CV^2$

Practice Questions

Q1 Define capacitance.

Q2 What is the relationship between charge, voltage and capacitance?

Q3 How would you find the energy stored on a capacitor from a graph of voltage against charge stored?

Exam Questions

Q1 Capacitors are used in cameras to give a brief flash of light when taking photographs.
Explain why a capacitor is a suitable component for this use. [2 marks]

Q2 A 500 mF capacitor is fully charged up from a 12 V supply.
The energy stored by a capacitor is given by the equation $E = \frac{1}{2}CV^2$.

(a) Choose the value from the list which gives the total energy stored by the capacitor:
    36 J   36 kJ   74 J   74 kJ [1 mark]

(b) Calculate the charge stored by the capacitor. [2 marks]

Capacitance — fun, it's not...

Capacitors are really useful in the real world. Pick an appliance, any appliance, and it'll probably have a capacitor or several. If I'm being honest though, the only saving grace of these pages for me is that they're not especially hard...

Charging and Discharging

Charging and discharging — sounds painful...

You can **Charge** a **Capacitor** by Connecting it to a **Battery**

1) When a capacitor is connected to a **battery**, a **current** flows in the circuit until the capacitor is **fully charged**, then **stops**.

2) The electrons flow onto the plate connected to the **negative terminal** of the battery, so a **negative charge** builds up.

3) This build-up of negative charge **repels** electrons off the plate connected to the **positive terminal** of the battery, making that plate positive. These electrons are attracted to the positive terminal of the battery.

4) An **equal** but **opposite** charge builds up on each plate, causing a **potential difference** between the plates.

Remember that **no charge** can flow **between** the plates because they're **separated** by an **insulator** (dielectric).

5) Initially the **current** through the circuit is **high**. But, as **charge** builds up on the plates, **electrostatic repulsion** makes it **harder** and **harder** for more electrons to be deposited. When the p.d. across the **capacitor** is equal to the p.d. across the **battery**, the **current** falls to **zero**. The capacitor is **fully charged**.

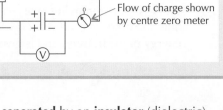

Flow of charge shown by centre zero meter

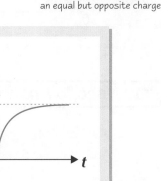

an equal but opposite charge

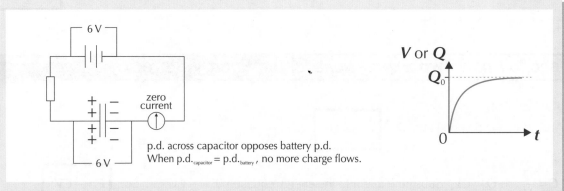

p.d. across capacitor opposes battery p.d.
When p.d.$_{capacitor}$ = p.d.$_{battery}$, no more charge flows.

To **Discharge** a Capacitor, **Take Out** the **Battery** and **Reconnect** the **Circuit**

1) When a **charged capacitor** is connected across a **resistor**, the p.d. drives a **current** through the circuit.

2) This current flows in the **opposite direction** from the **charging current**.

3) The capacitor is **fully discharged** when the **p.d.** across the plates and the **current** in the circuit are both **zero**.

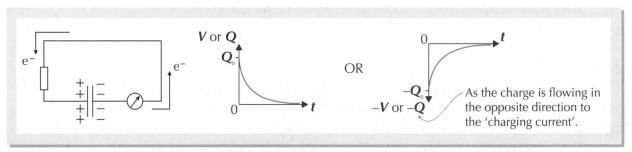

As the charge is flowing in the opposite direction to the 'charging current'.

The **Time Taken** to **Charge** or **Discharge** Depends on **Two Factors**

The **time** it takes to charge up or discharge a capacitor depends on:

1) The **capacitance** of the capacitor (**C**). This affects the amount of **charge** that can be transferred at a given **voltage**.

2) The **resistance** of the circuit (**R**). This affects the **current** in the circuit.

Charging and Discharging

Discharge Rate is Proportional to the Charge Remaining

If you measure the amount of **charge remaining** on the plates of a capacitor while it is **discharging**, you'll get a **graph** like the one on the **right**. The amount of charge **initially falls quickly**, but the rate **slows** as the amount of **charge decreases** — the **rate of discharge** is **proportional** to the **charge remaining**.

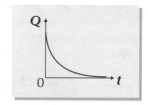

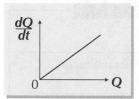

You can show this **relationship** by drawing a **graph** of the rate of discharge (**dQ/dt**) against the charge remaining (**Q**) — you get a lovely straight line through the origin (see left).

It also means that you can write an **equation** for the **rate of discharge** of a capacitor:

The exact solution of this equation is the exponential equation given below.

$$\frac{dQ}{dt} = -\frac{Q}{RC}$$

where **Q** is the **charge remaining** (C), **R** is the **resistance** of the circuit (Ω) and **C** is the **capacitance** of the capacitor (F).

The Charge on a Capacitor Decreases Exponentially

1) When a capacitor is **discharging**, the amount of **charge** left on the plates falls **exponentially with time**.
2) That means it always takes the **same length of time** for the charge to **halve**, no matter **how much charge** you start with — like radioactive decay (see p. 5).

The charge left on the plates of a capacitor discharging from full is given by the equation:

$$Q = Q_0 e^{-\frac{t}{RC}}$$

where Q_0 is the charge of the capacitor when it's fully charged.

The graphs of **V** against **t** and **I** against **t** for charging and discharging are also exponential.

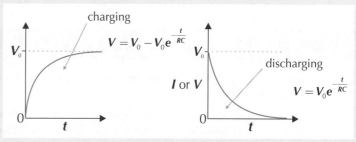

$$V = V_0 - V_0 e^{-\frac{t}{RC}}$$
charging

$$V = V_0 e^{-\frac{t}{RC}}$$
discharging

The current-time graph for a charging capacitor is just the same as the one for a discharging capacitor.

Time Constant τ = RC

τ is the Greek letter 'tau'

If $t = \tau = RC$ is put into the equation above, then $Q = Q_0 e^{-1}$.

So when $t = \tau$: $\frac{Q}{Q_0} = \frac{1}{e}$, where $\frac{1}{e} \approx \frac{1}{2.718} \approx 0.37$.

1) So τ, the **time constant**, is the time taken for the charge on a discharging capacitor (Q) to **fall** to **37%** of Q_0, or for the charge of a charging capacitor to **rise** to **63%** of Q_0.
2) The **larger** the **resistance** in series with the capacitor, the **longer it takes** to charge or discharge.

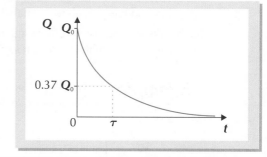

Practice Questions

Q1 Sketch graphs to show the variation of p.d. across the plates of a capacitor with time for:
a) charging a capacitor, b) discharging a capacitor.

Exam Question

Q1 A fully charged 250 µF capacitor is discharged through a 1 kΩ resistor.
(a) Calculate the time taken for the charge on the capacitor to fall to 37% of its original value. [2 marks]
(b) Calculate the percentage of the total charge remaining on the capacitor after 0.7s. [3 marks]

An analogy — consider the lowly bike pump...

One way to think of the charging process is like pumping air into a bike tyre. To start with, the air goes in easily, but as the tyre pressure increases, it gets harder and harder to squeeze more air in. The analogy works just as well for discharging...

Modelling Decay

Who'd have thought that capacitors and radioactive isotopes could have so much in common? Read on...

Capacitors *and* Radioactive Isotopes *Have Similar* Decay Equations

Radioactive isotopes (pages 4-5) might seem very different from **capacitors** in R-C circuits (pages 8-9), but the **decay models** for them are very similar. This **table** shows the **similarities** and **differences** between the models.

	Discharging Capacitors	Radioactive Isotopes
1)	Decay equation is $Q = Q_0 e^{-t/RC}$.	Decay equation is $N = N_0 e^{-\lambda t}$.
2)	The **quantity** that decays is Q, the amount of charge left on the plates of the capacitor.	The **quantity** that decays is N, the number of unstable nuclei remaining.
3)	**Initially**, the charge on the plates is Q_0.	**Initially**, the number of nuclei is N_0.
4)	It takes RC seconds for the amount of charge remaining to fall to **37% of its initial value**.	It takes $1/\lambda$ seconds for the number of nuclei remaining to fall to **37% of the initial value**.
5)	The time taken for the amount of charge left to decay by half (the **half-life**) is $t_{\frac{1}{2}} = \ln 2 \times RC$.	The time taken for the number of nuclei to decay by half (the **half-life**) is $t_{\frac{1}{2}} = \ln 2 / \lambda$.

You Can Use a Logarithmic Graph to Find the Decay Constant and Half-life

1) If you plot a **graph** of the **number of undecayed nuclei** of a radioactive sample against **time**, you get an **exponential curve** like the one on page 5.

2) But, if you plot the **natural log** (ln) of the number of **undecayed nuclei** against **time**, you get a **straight line**. To find the **natural log** of a number, just use the **ln button** on your calculator.

3) You get a straight line because the **decay equation**, $N = N_0 e^{-\lambda t}$, can be **rearranged**, via the mystical wonder of **logs**, to the **general form** of a **straight line** — $y = mx + c$.

$$N = N_0 e^{-\lambda t} \implies \ln(N) = -\lambda t + \ln(N_0)$$
$$y = mx + c$$

4) The **gradient** of the line is $-\lambda$, the **decay constant**. From this you can **calculate** the **half-life** of the sample.

5) This works for graphs of **activity** against **time** too — as long as you remember to **subtract** the **background activity** first.

If you understand logs you can work this out for yourself — if not, you'll just have to believe me. There's more stuff about logs on page 78.

Example   The graphs below show how the number of undecayed nuclei of a radioactive isotope decreases over time. Calculate the half-life of the isotope using both graphs.

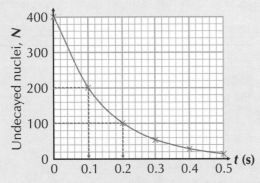

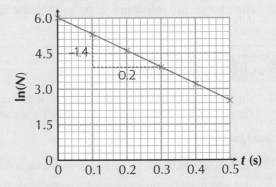

The first graph shows that it takes 0.1 s for the number of undecayed nuclei to fall from 200 to 100 (i.e. to halve). So, the **half-life is 0.1 s**.

The **gradient** of this graph is $-1.4 \div 0.2 = -7$, so the **decay constant**, $\lambda = 7$. Substitute this into the equation for half-life: $t_{\frac{1}{2}} = \ln(2) \div 7 = $ **0.1 s**.

You can use the **same method** with $Q = Q_0 e^{-\frac{t}{RC}}$ to find the **time constant** for an **R-C circuit** — the gradient is $-\dfrac{1}{RC}$.

Modelling Decay

You Can **Solve Differential Equations** Using **Iterative Methods**

1) $\dfrac{dQ}{dt} = -\dfrac{Q}{RC}$ is a **differential equation** with an **exact solution**: $Q = Q_0 e^{-\frac{t}{RC}}$.

2) Many differential equations **don't have** exact solutions, so scientists use **iterative numerical methods** to solve them. Iterative methods work for **every type** of differential equation and are easy to do with a **computer**. However, the answers are only **approximate**.

3) The method below can be used for **capacitor decay**, **radioactive decay** or **any** other model where the **rate of change** is **related** to the quantity that's changing — you just need to substitute the **relevant equation**.

1) Start with the **initial value of Q** (or use the value you have worked out in step 5).

2) **Substitute** the value of **Q** into the equation to work out the value of **dQ/dt**.

3) Increase the **time** by a small **interval** (the smaller the interval, the more accurate your answer).

4) Estimate the **change in Q** over this time interval by multiplying **dQ/dt** by the time interval.

5) Find the value of **Q after** the time interval by **adding the change** to the old value of **Q**, then go back to step 1 and **repeat the process** until you get to the time that you want.

Example

A 150 μF capacitor is charged until it stores 1.8 C of charge, then discharged through a resistance of 40 kΩ. Use an iterative method to find the charge remaining on the capacitor after 0.4 s of discharging. Use a time interval of 0.2 s for each iteration.

A table is useful for answering this type of question — start by filling in the time column and the first value of **Q**. Then divide **−Q** by **RC** to find the first value of **dQ/dt** and add this to the table. Next, multiply **dQ/dt** by the time interval (0.2 s), to find **ΔQ**. Add **ΔQ** to the initial value of **Q** to find the value of **Q** after the time interval. Copy the new value of **Q** into the next row and repeat the process to complete the rest of the row. You now have a value for **Q** after the time the question asks for (0.4 s) — so the answer is **1.682 C**.

$$\frac{-1.8}{40 \times 10^3 \times 150 \times 10^{-6}} \quad -0.3 \times 0.2 \quad 1.8 + (-0.06)$$

Time (s)	Q (C)	dQ/dt (Cs⁻¹)	ΔQ (C)	Q (C)
0	1.8	−0.3	−0.06	1.74
0.2	1.74	−0.29	−0.058	1.682
0.4	1.682			

Practice Questions

Q1 What are the decay equations for the discharge of a capacitor and a radioactive isotope?

Q2 How can you use a logarithmic graph to find: a) the decay constant, b) the time constant?

Exam Questions

Q1 A teacher set up an R-C circuit to model the radioactive decay of an isotope of radon gas. The values of R and C were chosen so that the charge and resistance were related to the decay constant of the radon by $RC = \lambda^{-1}$.

The capacitance of the capacitor was 500 μF and the resistance of the circuit was 144 kΩ.
Find the value of the decay constant of the radon. [2 marks]

Q2 A 10 μF capacitor is discharged through a resistance of 200 kΩ. The capacitor initially stores 5.0×10^{-2} C.
Use a table with the headings below to work out the charge after 2.0 s.
Use a time interval of 0.5s for each iteration. [5 marks]

Time (s)	Charge (C)	dQ/dt (Cs⁻¹)	ΔQ (C)	New Charge (C)

Modelling decay — an ageing catwalk queen...

The point of these two pages is for you to learn how to solve decay equations (if you hadn't already worked that out).

Simple Harmonic Motion

Radioactive decay to capacitors to... simple harmonic motion? Well, of course, they're all examples of modelling.

SHM is Defined in terms of Acceleration and Displacement

The **motion** of some **oscillating systems**, e.g. a **pendulum**, can be **modelled** by **simple harmonic motion** (SHM).

1) An object moving with **simple harmonic motion oscillates** to and fro, either side of a **midpoint**.

2) The distance of the object from the midpoint is called its **displacement**.

3) There is always a **restoring force** pulling or pushing the object back **towards** the **midpoint**.

4) The **size** of the **restoring force** depends on the **displacement**, and the force makes the object **accelerate** towards the midpoint:

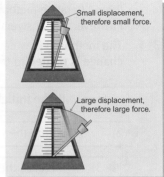

Midpoint

Small displacement, therefore small force.

Large displacement, therefore large force.

> **SHM:** an oscillation in which the **acceleration** of an object is **directly proportional** to its **displacement** from the **midpoint**, and is directed **towards the midpoint**.

The Restoring Force makes the Object Exchange PE and KE

1) The **type** of **potential energy** (PE) depends on **what it is** that's providing the **restoring force**. This will be **gravitational PE** for pendulums and **elastic PE** (elastic stored energy) for masses on springs.

2) As the object moves **towards the midpoint**, the restoring force **does work** on the object and so **transfers** some **PE** to **KE**. When the object is moving **away from the midpoint**, all that KE is transferred **back to PE** again.

3) At the **midpoint**, the object's **PE** is **zero** and its **KE** is **maximum**.

4) At the **maximum displacement** (the **amplitude**) on both sides of the midpoint, the object's **KE** is **zero** and its **PE** is **maximum**.

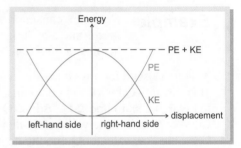

Energy

PE + KE

PE

KE

displacement

left-hand side   right-hand side

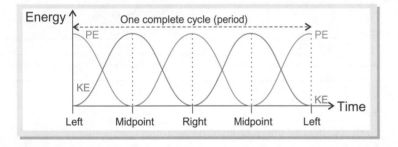

Energy

One complete cycle (period)

PE

PE

KE

KE

Time

Left   Midpoint   Right   Midpoint   Left

5) The **sum** of the **potential** and **kinetic** energy is called the **mechanical energy** and **stays constant** (as long as the motion isn't damped — see p. 16-17).

6) The **energy transfer** for one complete cycle of oscillation (see graph) is: PE to KE to PE to KE to PE … and then the process repeats…

You can Draw Graphs to Show Displacement, Velocity and Acceleration

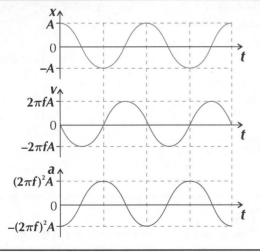

Displacement, x, varies as a cosine or sine wave with a maximum value, A (the amplitude). It's a cosine wave (as shown) when the stopwatch is started with the mass at the maximum displacement.

Velocity, v, is the gradient of the displacement-time graph (dx/dt). It has a maximum value of $(2\pi f)A$ (where f is the frequency of the oscillation) and is a quarter of a cycle in front of the displacement.

Acceleration, a, is the gradient of the velocity-time graph (d^2x/dt^2). It has a maximum value of $(2\pi f)^2A$, and is in antiphase with the displacement.

Simple Harmonic Motion

The **Frequency** and **Period** don't depend on the **Amplitude**

1) From **maximum positive displacement** (e.g. maximum displacement to the right) to **maximum negative displacement** (e.g. maximum displacement to the left) and **back again** is called a **cycle** of oscillation.

2) The **frequency**, *f*, of the SHM is the number of cycles per second (measured in Hz).

3) The **period**, *T*, is the **time** taken for a complete cycle (in seconds).

4) The relationship between **frequency** and **period** is given by the equation:
In other words, they're **inversely proportional** to each other.

$$f = \frac{1}{T}$$

> In SHM, the **frequency** and **period** are independent of the **amplitude** (i.e. constant for a given oscillation). So a pendulum clock will keep ticking in regular time intervals even if its swing becomes very small.

Learn the SHM Equations

1) According to the definition of SHM, the **acceleration**, *d²x/dt²*, is directly proportional to the **displacement**, *x*.

   The **constant of proportionality** depends on the **frequency**, and the acceleration is always in the **opposite direction** from the displacement (so there's a minus sign in the equation).

   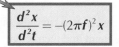
   $$\frac{d^2x}{d^2t} = -(2\pi f)^2 x$$

   This is another differential equation.

2) The **velocity** is **positive** when the object's moving in one direction, and **negative** when it's moving in the opposite direction. For example, a **pendulum's velocity** is **positive** when it's moving from **left to right** and **negative** when it's moving from **right to left**.

3) The **displacement** varies with time according to two equations, depending on **where** the object was when the timing was started — don't worry, they're really similar so you shouldn't have too much trouble learning both.

> For someone starting a stopwatch with a pendulum at **maximum displacement**:
> $$x = A\cos(2\pi ft)$$

> For someone releasing a pendulum but starting a stopwatch as the pendulum swings through **the midpoint**:
> $$x = A\sin(2\pi ft)$$

Practice Questions

Q1 Sketch a graph of how the velocity of an object oscillating with SHM varies with time.

Q2 State the equation linking frequency and period.

Q3 What is the special relationship between the acceleration and the displacement in SHM?

Q4 What are the two equations for the displacement of a simple harmonic oscillator?

Exam Questions

Q1 (a) Define *simple harmonic motion*. [2 marks]

   (b) Explain why the motion of a ball bouncing off the ground is not SHM. [1 mark]

Q2 A pendulum is pulled a distance 0.05 m from its midpoint and released.
   It oscillates with simple harmonic motion with a frequency of 1.5 Hz.

   (a) Choose a value from the list below which gives the maximum velocity of the pendulum.
      0.24 ms⁻¹   9.4 ms⁻¹   0.15 ms⁻¹   0.47 ms⁻¹ [1 mark]

   (b) Calculate its displacement 0.1 s after it is released. [2 marks]

"Simple" harmonic motion — hmmm, I'm not convinced...

The basic concept of SHM is simple enough (no pun intended). Make sure you can remember the shapes of all the graphs on page 12 and the equations from this page, then just get as much practice at using the equations as you can.

Simple Harmonic Oscillators

A *Mass* on a *Spring* is a *Simple Harmonic Oscillator (SHO)*

1) When the mass is **pushed to the left** or **pulled to the right** of the **equilibrium position**, there's a **force** exerted on it. The size of this force is:

$$F = kx$$

where **k** is the **spring constant** (stiffness) of the spring in Nm⁻¹ and **x** is the displacement in m.

2) After a bit of jiggery-pokery involving Newton's second law (**F = ma**) and some of the ideas on the previous page, you get the **formula for the period of a mass oscillating on a spring**:

$$T = 2\pi\sqrt{\frac{m}{k}}$$

where **T** = period of oscillation in seconds
m = mass in kg
k = spring constant in Nm⁻¹

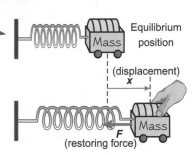

Equilibrium position

(displacement) **x**

F (restoring force)

You can check this result **EXPERIMENTALLY** by changing **one variable at a time** and seeing what happens.

Investigating the Mass-Spring System

spring constant **k**

mass **m**

position sensor

trolley

1) Attach a **trolley** between two **springs**, pull it to one side by a certain amount and then let go. The trolley will **oscillate** back and forth as the springs pull it in each direction.

2) Measure the **period, T**, by getting a computer to plot a **displacement-time graph**.

3) Change the **mass, m**, by loading the trolley with **masses** — don't forget to include the mass of the trolley in your calculations.

4) Change the **spring stiffness, k**, by using different combinations of springs.

5) Change the **amplitude, A**, by pulling the trolley across by different amounts.

6) You'll get the following **results**: (∝ *means "is proportional to"*)

   a) $T \propto \sqrt{m}$ so $T^2 \propto m$ b) $T \propto \sqrt{\frac{1}{k}}$ so $T^2 \propto \frac{1}{k}$ c) **T** doesn't depend on amplitude, **A**.

$k \rightarrow 2k \rightarrow 3k$

$\frac{1}{2}k$

$\frac{1}{3}k$

Compressing or *Stretching* a *Spring* Stores *Elastic Potential Energy*

When the mass is **pushed** or **pulled** away from the **equilibrium point**, the spring is **compressed** or **stretched** so stores **elastic potential energy**. You can find how much energy is stored by plotting a **force-extension graph** — the **area under the graph** is the **elastic potential energy**.

Or, you can work it out using:

$$E = \frac{1}{2}kx^2$$

this formula is the same as working out the area of a triangle with base **kx** and height **x**.

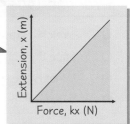

Extension, x (m)

Force, kx (N)

The *Simple Pendulum* is the *Classic Example* of an *SHO*

If you set up a simple pendulum attached to an angle sensor and computer like this — then change the length, **l**, the mass of the bob, **m**, and the amplitude, **A**, you get the following results:

a) $T \propto \sqrt{l}$, so $T^2 \propto l$ b) **T** does not depend on **m**   c) **T** does not depend on **A**

angle sensor

stiff rod

length, l

The **formula for the period of a pendulum** is:

$$T = 2\pi\sqrt{\frac{l}{g}}$$

This formula only works for small angles of oscillation — up to about 10° from the equilibrium point.

where **T** is the period of oscillation in seconds, **l** is the length of the pendulum in m, and **g** is the gravitational field strength in Nkg⁻¹.

Simple Harmonic Oscillators

The Force on an SHO is Proportional to its Displacement

The **definition** of **simple harmonic motion** (page 12) states that the **acceleration** of an SHO is **proportional** to its **displacement**. **Acceleration** is also **proportional** to **force** ($F = ma$ remember), so the **force** on an SHO must be **proportional** to its **displacement**.

Put all this together and you get: $F = ma = m\dfrac{d^2x}{dt^2}$ and $F = -kx$ (page 14), which means that $\boxed{\dfrac{d^2x}{dt^2} = -\dfrac{k}{m}x}$

You can use this differential equation to **estimate** the **displacement** of an SHO after a certain time using an **iterative method** like the one for capacitor and radioactive decay on page 11.

Example A 1 kg mass is attached to a mass-spring system with a spring constant of 15 Nm^{-1}. The mass is displaced by 0.1 m, then released and allowed to oscillate freely. Use an iterative method to find the displacement of the mass after 0.4 s. Use a time interval of 0.2 s for each iteration.

1) Draw a table like the one below and fill in the time column and the first values of **x** and **v**. The initial velocity is 0 ms^{-1} because the mass is held at a fixed displacement before being released.

2) Multiply **–k/m** by **x** to find the first value of **d²x/dt²** (acceleration) and add this to the table.

3) Multiply **d²x/dt²** by the time interval (0.2 s), to find the change in velocity, **Δv**, during the time interval. Add **Δv** to the initial value of **v** to find the velocity after the time interval.

4) Find **Δx** — the change in displacement during the time interval, by multiplying **v** by the time interval. Add this to the initial value of **x** to find the new value.

5) Copy the new values of **x** and **v** into the next row and repeat the process to complete the rest of the row.

$$-\frac{15}{1} \times 0.1 \qquad -1.5 \times 0.2 \qquad 0 + (-0.3) \qquad -0.3 \times 0.2 \qquad 0.1 + (-0.06)$$

Time (s)	x (m)	v (ms⁻¹)	d²x/dt² (ms⁻²)	Δv (ms⁻¹)	v (ms⁻¹)	Δx (m)	x (m)
0	0.1	0	–1.5	–0.3	–0.3	–0.06	0.04
0.2	0.04	–0.3	–0.6	–0.12	–0.42	–0.084	–0.044
0.4	–0.044						

6) You now have a value for **x** after the time the question asks for (0.4 s) — so the answer is **–0.044 m**.

Practice Questions

Q1 Write down the formulae for the period of a mass on a spring and the period of a pendulum.

Q2 Describe a method you could use to measure the period of an oscillator.

Q3 For a mass-spring system, what graphs could you plot to find out how the period depends on:
a) the mass, b) the spring constant, and c) the amplitude? What would they look like?

Exam Questions

Q1 A spring of original length 0.10 m is suspended from a stand and clamp.
A mass of 0.10 kg is attached to the bottom and the spring extends to a total length of 0.20 m.

(a) Calculate the spring constant of the spring in Nm^{-1}. ($g = 9.81$ Nkg^{-1}) The spring isn't moving at this point, so the forces on it must be balanced. [2 marks]

(b) The mass is pulled down a further 2 cm and then released.
The spring oscillates with simple harmonic motion, with a period of 0.63 s.
What mass would be needed to make the period of oscillation twice as long? [2 marks]

Q2 Two pendulums of different lengths were released from rest at the top of their swing.
It took exactly the same time for the shorter pendulum to make five complete oscillations
as it took the longer pendulum to make three complete oscillations.
The shorter pendulum had a length of 0.20 m. Show that the length of the longer one was 0.56 m. [3 marks]

Go on — SHO the examiners what you're made of...

The most important things to remember on these pages are those two period equations. You'll be given them in your exam, but you need to know what they mean and be happy using them.

Free and Forced Vibrations

Resonance... hmm... tricky little beast. Remember the Millennium Bridge, that standard-bearer of British engineering? The wibbles and wobbles were caused by resonance. How was it sorted out? By damping, which is coming up too.

Free Vibrations — *No Transfer* of *Energy* to or from the *Surroundings*

1) If you stretch and release a mass on a spring, it oscillates at its **natural frequency**.

2) If **no energy's transferred** to or from the surroundings, it will **keep** oscillating with the **same amplitude forever**.

3) In practice this **never happens**, but a spring vibrating in air is called a **free vibration** anyway.

> You need to know this formula for the **total energy** of a freely oscillating mass on a spring:
> $$E_{total} = \frac{1}{2}mv^2 + \frac{1}{2}kx^2 \text{ (in other words, K.E. + P.E.)}$$

Forced Vibrations happen when there's an External Driving Force

1) A system can be **forced** to vibrate by a periodic **external force**.

2) The frequency of this force is called the **driving frequency**.

Resonance happens when Driving Frequency = Natural Frequency

When the **driving frequency** approaches the **natural frequency**, the system gains more and more energy from the driving force and so vibrates with a **rapidly increasing amplitude**. When this happens the system is **resonating**.

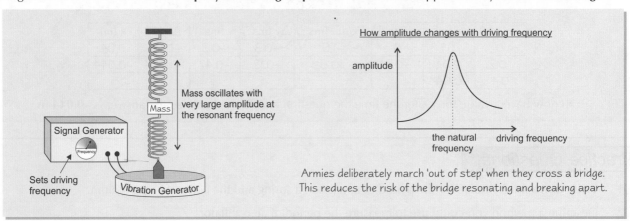

Signal Generator — Sets driving frequency — Vibration Generator — Mass oscillates with very large amplitude at the resonant frequency

How amplitude changes with driving frequency — amplitude — the natural frequency — driving frequency

Armies deliberately march 'out of step' when they cross a bridge. This reduces the risk of the bridge resonating and breaking apart.

Examples of resonance:

a) organ pipe

The column of air resonates, driven by the motion of air at the base.

b) swing

A swing resonates if it's driven by someone pushing it at its natural frequency.

c) glass smashing

A glass resonates when driven by a sound wave at the right frequency.

d) radio

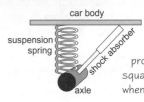

A radio is tuned so the electric circuit resonates at the same frequency as the radio station you want to listen to.

Damping happens when Energy is Lost to the Surroundings

1) In practice, **any** oscillating system **loses energy** to its surroundings.

2) This is usually down to **frictional forces** like air resistance.

3) These are called **damping forces**.

4) Systems are often **deliberately damped** to **stop** them oscillating or to **minimise** the effect of **resonance**.

car body — suspension spring — shock absorber — axle

Shock absorbers in a car suspension provide a damping force by squashing oil through a hole when compressed.

Free and Forced Vibrations

Different Amounts of Damping have Different Effects

1) The **degree** of damping can vary from **light** damping (where the damping force is small) to **overdamping**.

2) Damping **reduces** the **amplitude** of the oscillation over time. The **heavier** the damping, the **quicker** the amplitude is reduced to zero.

3) **Critical damping** reduces the amplitude (i.e. stops the system oscillating) in the **shortest possible time**.

4) Car **suspension systems** and moving coil **meters** are critically damped so that they **don't oscillate** but return to equilibrium as quickly as possible.

5) Systems with **even heavier damping** are **overdamped**. They take **longer** to return to equilibrium than a critically damped system.

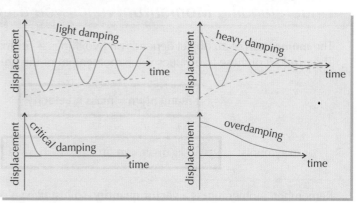

6) **Plastic deformation** of ductile materials **reduces** the **amplitude** of oscillations in the same way as damping. As the material changes shape, it **absorbs energy**, so the oscillation will become smaller.

Damping Affects Resonance too

1) **Lightly damped** systems have a **very sharp** resonance peak. Their amplitude only increases dramatically when the **driving frequency** is **very close** to the **natural frequency**.

2) **Heavily damped** systems have a **flatter response**. Their amplitude doesn't increase very much near the natural frequency and they aren't as **sensitive** to the driving frequency.

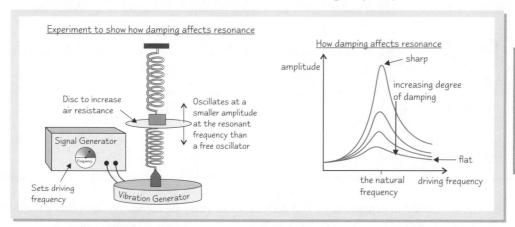

Structures are damped to avoid being damaged by resonance. Loudspeakers are also made to have as flat a response as possible so that they don't 'colour' the sound.

Practice Questions

Q1 What is a free vibration? What is a forced vibration?

Q2 Draw diagrams to show how a damped system oscillates with time when the system is lightly damped and when the system is critically damped.

Exam Questions

Q1 (a) Describe what resonance is and when it occurs. [2 marks]

(b) Draw a diagram to show how the amplitude of a lightly damped system varies with driving frequency. [2 marks]

(c) On the same diagram, show how the amplitude of the system varies with driving frequency when it is heavily damped. [1 mark]

Q2 Shock absorbers are used to critically damp the springs of a car's suspension system. Describe what happens when the suspension spring is compressed in terms of oscillations and equilibrium. [2 marks]

A2 Physics — it can really put a damper on your social life...

Resonance can be really useful (radios, oboes, swings — yay) or very, very bad...

Forces and Momentum

We're going to kick off with a couple of pages about linear momentum — that's momentum in a straight line (not a circle or anything complicated like that). You'll have met momentum before so it shouldn't be too hard.

Understanding **Momentum** helps you do **Calculations** on **Collisions**

The **momentum** of an object depends on two things — its **mass** and **velocity**.
The **product** of these two values is the momentum of the object.

$$\boxed{\textbf{momentum} = \textbf{mass} \times \textbf{velocity}}$$

or in symbols: $\boxed{\boldsymbol{p}\ (\text{in kg ms}^{-1}) = \boldsymbol{m}\ (\text{in kg}) \times \boldsymbol{v}\ (\text{in ms}^{-1})}$

Hands up if you want to learn some physics.

Remember, momentum is a **vector quantity**, so just like velocity, it has size and direction.

Momentum is always **Conserved**

1)   Assuming **no external forces** act, momentum is always **conserved**.

2)   This means the **total momentum** of two objects **before** they collide **equals** the total momentum **after** the collision.

3)   This is really handy for working out the **velocity** of objects after a collision (as you do...):

> *Example*   A skater of mass 75 kg and velocity 4 ms^{-1} collides with a stationary skater of mass 50 kg.
> The two skaters join together and move off in the same direction. Calculate their velocity after impact.
>
>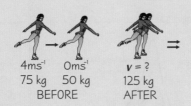
>
> 4ms^{-1} 0ms^{-1} v = ?
> 75 kg   50 kg   125 kg
> BEFORE   AFTER
>
> *Before you start a momentum calculation, always draw a quick sketch.*
>
> Momentum of skaters before = Momentum of skaters after
> $(75 \times 4) + (50 \times 0) = 125v$
> $300 = 125v$
> So $v = 2.4$ ms^{-1}

4)   The same principle can be applied in **explosions**. E.g. if you fire an **air rifle**, the **forward momentum** gained by the pellet **equals** the **backward momentum** of the rifle, and you feel the rifle recoiling into your shoulder.

> *Example*   A bullet of mass 0.005 kg is shot from a rifle at a speed of 200 ms^{-1}.
> The rifle has a mass of 4 kg. Calculate the velocity at which the rifle recoils.
>
>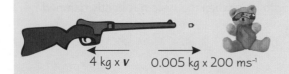
>
> 4 kg x v 0.005 kg x 200 ms^{-1}
>
> Momentum before explosion = Momentum after explosion
> $0 = (0.005 \times 200) + (4 \times v)$
> $0 = 1 + 4v$
> $v = -0.25$ ms^{-1}

5)   In reality, collisions usually happen in more than one dimension.
Don't worry though — you'll only be given problems to solve in **one dimension**.

Forces and Momentum

Rocket Propulsion can be Explained by Momentum

For a **rocket** to be **propelled forward** it must expel **exhaust gases**. The momentum of the rocket in the forward direction is **equal** to the momentum of the exhaust gases in the backward direction.

Example

A rocket of mass 500 kg is completely stationary in an area of space a long way from any gravitational fields. It starts its engines. The rocket ejects 2.0 kg of gas per second at a speed of 1000 ms^{-1}. Calculate the velocity of the rocket after 1 second.
(For the purpose of this example ignore the loss of mass due to fuel use.)

2 kg
1000 ms^{-1}

500 kg
v = ?

The total momentum before rocket fires = 0 kg ms^{-1} as initially the rocket is stationary; the total momentum after rocket fires must also = 0 kg ms^{-1}.

Total momentum after rocket fires = $(500 \times v) + (2 \times 1000) = 0$

$$500v = -2000$$
$$v = -4 \text{ ms}^{-1}$$

The minus sign shows that the rocket moves in the opposite direction to the exhaust gases.

Later, the rocket prepares to dock with a space station. The rocket is travelling at a constant speed of 0.5 ms^{-1} relative to the space station, which has a mass of 3500 kg. Calculate the change in the velocity of the space station when the rocket docks with it.

500 kg
v = 0.5 ms^{-1}

Total momentum before docking = Total momentum after docking

$$500 \times 0.5 = (3500 + 500)v$$
$$v = 250 \div 4000$$
$$v = 0.0625 \text{ ms}^{-1}$$
$$\text{change in velocity} = 0.0625 - 0 = 0.0625 \text{ ms}^{-1}$$

In practice, the space station and rocket would both be travelling at several thousand metres per second — what's important though is the difference in speed between them, i.e. their relative speeds.

Practice Questions

Q1 What is the equation for momentum?

Q2 Is momentum a vector or scalar quantity? What does this mean?

Q3 Explain how rocket propulsion is an example of the conservation of momentum.

Q4 Give another example of conservation of momentum in practice.

Exam Questions

Q1 A ball of mass 0.6 kg moving to the right at 5 ms^{-1} collides with a larger stationary ball of mass 2 kg.
The smaller ball rebounds, moving to the left at 2.4 ms^{-1}.
Calculate the velocity of the larger ball immediately after the collision. [3 marks]

Q2 A toy train of mass 0.7 kg, travelling at 0.3 ms^{-1}, collides with a stationary toy carriage of mass 0.4 kg.
The two toys couple together. Show that their new velocity is 0.19 ms^{-1}. [3 marks]

Q3 A spacecraft of mass 4200 kg is completely stationary in space. It is a long way from any gravitational fields. The spacecraft starts its engines and starts moving. After 1 second it is travelling at 2.5 ms^{-1}.
Calculate the momentum of the gas the spacecraft must eject to reach this velocity.
(You should ignore the spacecraft's loss of mass due to fuel use.) [2 marks]

<u>Momentum will never be an endangered species — it's always conserved...</u>

So, guess what — momentum is conserved in collisions. If you forget that you'll really scupper your chances of getting lots of marks from momentum calculations. So remember, momentum is mass times velocity, and momentum is conserved. See, it's easy — momentum is your friend.

Newton's Laws of Motion

You did most of this at GCSE, but that doesn't mean you can just skip over it now. You'll be kicking yourself if you forget this stuff in the exam — easy marks...

Newton's **1st Law** says that a **Force** is Needed to Change Velocity

1) **Newton's 1st law of motion** states the **velocity** of an object will **not change** unless a **resultant force** acts on it.

2) In plain English this means a body will remain at rest or moving in a **straight line** at a **constant speed**, unless acted on by a **resultant force**.

An apple sitting on a table won't go anywhere because the **forces** on it are **balanced**.

reaction (**R**)   =   **weight** (mg)

(force of table   (force of gravity
pushing apple up)   pulling apple down)

3) If the forces **aren't balanced**, the **overall resultant force** will cause the body to **accelerate** — if you gave the apple above a shove, there'd be a resultant force acting on it and it would roll off the table. Acceleration can mean a change in **direction**, or **speed**, or both. (See Newton's 2nd law, below.)

Newton's **2nd Law** says that Force is the **Rate of Change in Momentum**

*"The **rate of change of momentum** of an object is **directly proportional** to the **resultant force** which acts on the object."*   so $F = \dfrac{\Delta mv}{\Delta t}$

If mass is constant, this can be written as the well-known equation:

resultant force (F) = mass (m) × acceleration (a)

Learn this — it crops up all over the place in A2 Physics. And learn what it means too:

1) It says that the **more force** you have acting on a certain mass, the **more acceleration** you get.

2) It says that for a given force the **more mass** you have, the **less acceleration** you get.

3) A **small force** acting for a **long time** can cause the **same change in momentum** as a **large force** acting for a **short time**.

REMEMBER:
1) The **resultant force** is the **vector sum** of all the forces.
2) The force is **always** measured in **newtons**. Always.
3) The **mass** is always measured in **kilograms**.
4) a is the **acceleration** of the object as a result of F. It's **always** measured in **metres per second per second** (ms⁻²)
5) The **acceleration** is always in the **same direction** as the **resultant force**.

For example, a toy car with a mass of 1 kg, travelling at 5 ms⁻¹, hits a wall and stops in a time of 0.5 seconds.

The average force on the car is: $F = \dfrac{mv - mu}{t} = \dfrac{(1 \times 5) - (1 \times 0)}{0.5} = 10\,N$

But if the time of impact is doubled to 1 second, the force on the car is halved.

This is the idea behind car crumple zones, which increase the time of an impact to reduce the force on the passengers.

F = ma is a **Special Case** of Newton's 2nd Law

Newton's 2nd law says that if the **mass** of an object is **constant**, then the **bigger** the **force** acting on it, the **greater** its **acceleration** — i.e. **F = ma**. But, if the **mass** of the object is **changing** — e.g. if it is accelerating at close to the **speed of light** — then you **can't** use **F = ma**.

Don't worry though — **Newton's 2nd law still applies**, it's just that the 'rate of **change of momentum**' bit refers to a **change in mass** and velocity.

Daisy was always being told that she was a special case.

Newton's Laws of Motion

Newton's **3rd Law** is a **Consequence** of the **Conservation of Momentum**

There are a few different ways of stating Newton's 3rd law, but the clearest way is:

> **If an object A EXERTS a FORCE on object B, then object B exerts AN EQUAL BUT OPPOSITE FORCE on object A.**

You'll also hear the law as "every action has an equal and opposite reaction". But this confuses people who wrongly think the forces are both applied to the same object. (If that were the case, you'd get a resultant force of zero and nothing would ever move anywhere...)

The two forces actually represent the **same interaction**, just seen from two **different perspectives**:

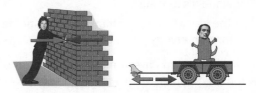

1) If you **push against a wall**, the wall will **push back** against you, **just as hard**. As soon as you stop pushing, so does the wall. Amazing...

2) If you **pull a cart**, whatever force **you exert** on the rope, the rope exerts the **exact opposite** pull on you.

3) When you go **swimming**, you push **back** against the water with your arms and legs, and the water pushes you **forwards** with an equal-sized force. So, the **backward momentum** of the water is equal to your **forward momentum**.

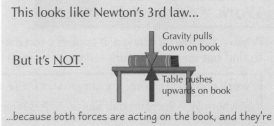

This looks like Newton's 3rd law...

Gravity pulls down on book

But it's <u>NOT</u>.

Table pushes upwards on book

...because both forces are acting on the book, and they're not of the same type. This is two separate interactions. The forces are equal and opposite, resulting in zero acceleration, so this is showing Newton's 1st law.

Newton's 3rd law applies in **all situations** and to all **types of force**. But the pairs of forces are always the **same type**, e.g. both gravitational or both electrical.

Newton's 3rd law is a consequence of the **conservation of momentum** (page 18). A **resultant force** acting means a change in **mass** or **acceleration** ($F = ma$) — which means a **change in momentum**. Momentum is always **conserved**, so whenever one object exerts a force on another (and changes its momentum), the second object must exert an equal-sized force back on the first object so that the overall change in momentum is zero.

Practice Questions

Q1 State Newton's 1st, 2nd and 3rd laws of motion, and explain what they mean.

Q2 Give an example of a situation where you couldn't use F = ma. Why wouldn't the equation apply?

Q3 Sketch a force diagram of a book resting on a table to illustrate Newton's 3rd law.

Exam Questions

Q1 A parachutist with a mass of 78 kg jumps out of a plane. As she falls, she accelerates.

(a) What is the initial vertical force on the parachutist? Use $g = 9.81$ ms^{-2}. [1 mark]

(b) After a time, the parachutist reaches terminal velocity and stops accelerating. Use Newton's 1st law to explain why the resultant force on the parachutist is zero at this point. [2 marks]

Q2 A 250 kg boat is moving across a river. The engines provide a force of 500 N at right angles to the flow of the river, and the boat experiences a drag of 100 N in the opposite direction. The force on the boat due to the flow of the river is 300 N. Show that the magnitude of the acceleration of the boat is 2 ms^{-2}. [4 marks]

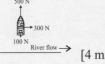

500 N

300 N

100 N

River flow →

Newton's three incredibly important laws of motion...

These equations may not really fill you with a huge amount of excitement (and I hardly blame you if they don't)... but it was pretty fantastic at the time — suddenly people actually understood how forces work, and how they affect motion. I mean arguably it was one of the most important scientific discoveries ever...

Work and Energy

As everyone knows, work in Physics isn't like normal work. It's harder. Work also has a specific meaning that's to do with movement and forces. You'll have seen this at GCSE — it just comes up in more detail for A2.

Work is done whenever Energy is Transferred

This table gives you some examples of **work being done** and the **energy changes** that happen.

1) Usually you need a force to move something because you're having to **overcome another force**.

2) The thing being moved has **kinetic energy** while it's **moving**.

3) The kinetic energy is transferred to **another form of energy** when the movement stops.

ACTIVITY	WORK DONE AGAINST	FINAL ENERGY FORM
Lifting up a box.	gravity	gravitational potential energy
Pushing a chair across a level floor.	friction	heat (thermal)
Pushing two magnetic north poles together.	magnetic force	magnetic energy
Stretching a spring.	stiffness of spring	elastic potential energy

The word **'work'** in Physics means the **amount of energy transferred** from one form to another when a force causes a movement of some sort.

Work = Force × Distance

When a car tows a caravan, it applies a force to the caravan to move it.
To **find out** how much **work** is **done**, you need to use the **equation**:

> **work done (W) = force causing motion (F) × distance moved (s)**
> ...where **W** is measured in joules (J), **F** is measured in newtons (N) and **s** is measured in metres (m).

Points to remember:

1) **Work** is the **energy** that's been **changed** from one form to another — it's not necessarily the **total** energy. E.g. moving a book from a low shelf to a higher one will increase its gravitational potential energy, but it had some potential energy to start with. Here, the **work done** would be the **increase** in potential energy, **not the total** potential energy.

2) Remember, the distance needs to be measured in metres — if you have **distance in centimetres or kilometres**, you need to **convert** to metres first.

3) The force **F** will be a **fixed** value in any calculations, either because it's **constant** or because it's the **average** force.

4) The equation assumes that the **direction of the force** is the **same** as the **direction of movement**.

5) The equation gives you the **definition** of the joule (symbol J):
'one joule is the work done when a force of 1 newton moves an object through a distance of 1 metre'

6) If you plotted a graph of force (**F**) against distance moved (**s**), the **area under the graph** would equal the work done.

The Force isn't always in the Same Direction as the Movement

Sometimes the **direction of movement** is **different** from the **direction of the force**.

Example

direction of force on sledge

rosebud

direction of motion

1) To **calculate the work done** in a situation like the one in the diagram, you need to consider the **horizontal** and **vertical components** of the **force**.

2) The only **movement** is in the **horizontal** direction. This means the **vertical force** is not causing any motion (and hence not doing any work) — it's just **balancing** out some of the **weight**, meaning there's a **smaller reaction force**.

3) The horizontal force is causing the motion — so to **calculate** the **work done**, this is the **only force** you need to consider. Which means we get:

$$W = Fs \cos \theta$$

Where θ is the **angle** between the **direction of the force** and the **direction of motion**.

F
θ
$F\cos\theta$
→ Direction of motion

The equation shows that if the **force is perpendicular** to the **direction of motion**, then $\cos \theta = 0$, so **no work will be done**. This is why you can **ignore** the component of the force at **right angles** to the movement — it does **no work**.

Work and Energy

Learn the **Principle** of **Conservation** of **Energy**

The **principle of conservation of energy** says that:

> Energy **cannot be created** or **destroyed**. Energy **can be transferred** from one form to another but the total amount of energy in a closed system will not change.

Example

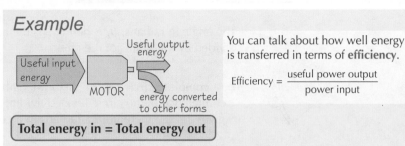

Useful input energy → MOTOR → Useful output energy

energy converted to other forms

Total energy in = Total energy out

You can talk about how well energy is transferred in terms of **efficiency**.

$$\text{Efficiency} = \frac{\text{useful power output}}{\text{power input}}$$

You need it for **Questions** about **Kinetic and Potential Energy**

The principle of conservation of energy nearly always comes up when you're doing questions about changes between kinetic and potential energy. Why — because energy is only ever exchanged from one form to another, not destroyed.

A quick reminder:

1) **Kinetic energy** is the energy of anything **moving**. There are two equations for kinetic energy that you need to know — one in terms of **mass** and **velocity**, and one in terms of **momentum**.

2) There are **different types of potential energy** — e.g. gravitational and elastic.

3) **Gravitational potential energy** is the energy something gains if you lift it up. It depends on the **mass** of the object, the **height** it is lifted and the value of **g**, the **gravitational field strength** ($9.81 \, \text{Nkg}^{-1}$ on Earth).

4) **Elastic potential energy** (elastic stored energy) is the energy you get in, say, a stretched rubber band or spring. You can find it by plotting a **force-extension graph** — the area under the graph is the elastic potential energy. Or, you can work it out with the equation shown — **x** is the extension of the spring and **k** is the stiffness constant — see p14-15.

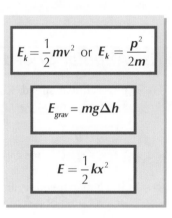

$$E_k = \frac{1}{2}mv^2 \text{ or } E_k = \frac{p^2}{2m}$$

$$E_{grav} = mg\Delta h$$

$$E = \frac{1}{2}kx^2$$

Practice Questions

Q1 Write down the equation used to calculate work if the force and motion are in the same direction.

Q2 Write down the equation for work if the force is at an angle to the direction of motion.

Q3 When will a force not do any work on a moving object?

Q4 Give two different equations for kinetic energy.

Exam Questions

Q1 A traditional narrowboat is drawn by a horse walking along a towpath. The horse pulls the boat at a constant speed between two locks which are 1500 m apart. The tension in the rope is 100 N at 40° to the direction of motion. Choose the value from the list below which gives the amount of work done on the boat to the nearest kJ. Show your working.

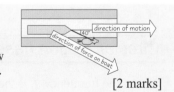

   115 kJ   125 kJ   135 kJ   145 kJ   [2 marks]

Q2 A motor is used to lift a 20 kg load a height of 3 m. (Take $g = 9.81 \, \text{Nkg}^{-1}$.)

   (a) Calculate the work done in lifting the load.   [2 marks]

   (b) The speed of the load during the lift is $0.25 \, \text{ms}^{-1}$. Calculate the power delivered by the motor.   [2 marks]

*Work, work, work — when will it all end..**

So work is the amount of energy needed for a force to move something a certain distance — easy. All you need to do is learn the equations and what to do when the force and movement are in different directions, and you'll be fine... * Answer: page 79

Circular Motion

*It's probably worth putting a bookmark in here — this stuff is needed **all over** the place.*

Angles can be Expressed in Radians

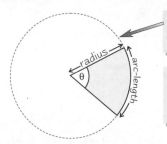

The angle in **radians**, θ, is defined as the **arc-length** divided by the radius of the circle.

For a **complete circle** (360°), the arc-length is just the circumference of the circle ($2\pi r$). Dividing this by the radius (r) gives 2π. So there are 2π radians in a complete circle.

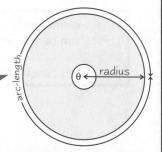

Some common angles:

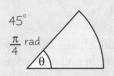

45°
$\dfrac{\pi}{4}$ rad

90°
$\dfrac{\pi}{2}$ rad

180°
π rad

angle in radians $= \dfrac{2\pi}{360} \times$ angle in degrees

1 radian is about 57°

The Angular Speed is the Angle an Object Rotates Through per Second

1) Just as **linear speed**, v, is defined as distance ÷ time, the **angular speed**, ω, is defined as **angle ÷ time**. The unit is rad s^{-1} — radians per second.

$$\omega = \frac{\theta}{t}$$

ω = angular speed (rad s^{-1}) — the symbol for angular speed is the little Greek 'omega', not a w.
θ = angle (radians) turned through in a time, t (seconds)

2) The **linear speed**, v, and **angular speed**, ω, of a rotating object are linked by the equation:

$$v = r\omega$$

v = linear speed (ms^{-1}), r = radius of the circle (m),
ω = angular speed (rad s^{-1})

Example — Beam of Particles in a Cyclotron

(See page 66)

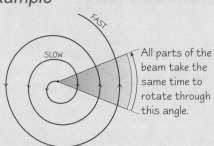

FAST
SLOW
All parts of the beam take the same time to rotate through this angle.

1) Different parts of the particle beam are rotating at **different linear speeds**, v. (The linear speed is sometimes called **tangential velocity**.)

2) But all the parts **rotate** through the **same angle** in the **same time** — so they have the **same angular speed**.

Circular Motion has a Frequency and Period

1) The frequency, f, is the number of complete **revolutions per second** (rev s^{-1} or hertz, Hz).

2) The period, T, is the **time taken** for a complete revolution (in seconds).

3) Frequency and period are **linked** by the equation:

$$f = \frac{1}{T}$$

f = frequency in rev s^{-1}, T = period in s

4) For a complete circle, an object turns through 2π radians in a time T, so frequency and period are related to ω by:

$$\omega = 2\pi f \quad \text{and} \quad \omega = \frac{2\pi}{T}$$

f = frequency in rev s^{-1}, T = period in s, ω = angular speed in rad s^{-1}

Circular Motion

Objects Travelling in Circles are *Accelerating* since their *Velocity is Changing*

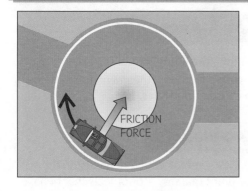

1) Even if the car shown is going at a **constant speed**, its **velocity** is changing since its **direction** is changing.

2) Since acceleration is defined as the **rate of change of velocity**, the car is accelerating even though it isn't going any faster.

3) This acceleration is called the **centripetal acceleration** and is always directed towards the **centre of the circle**.

There are two formulas for centripetal acceleration:

$$a = \frac{v^2}{r} \quad \text{and} \quad a = \omega^2 r$$

a = centripetal acceleration in ms^{-2}
v = linear speed in ms^{-1}
ω = angular speed in rad s^{-1}
r = radius in m

The *Centripetal Acceleration* is produced by a *Centripetal Force*

From Newton's laws, if there's a **centripetal acceleration**, there must be a **centripetal force** acting towards the **centre of the circle**.

Since **F = ma**, the centripetal force must be:

$$F = \frac{mv^2}{r} \quad \text{and} \quad F = m\omega^2 r$$

The centripetal force is what keeps the object moving in a circle — remove the force and the object would fly off at a tangent.

Men cowered from the force of the centripede.

Practice Questions

Q1 How many radians are there in a complete circle?

Q2 How is angular speed defined and what is the relationship between angular speed and linear speed?

Q3 Define the period and frequency of circular motion. What is the relationship between period and angular speed?

Q4 In which direction does the centripetal force act, and what happens when this force is removed?

Exam Questions

Q1 The Earth orbits the Sun with an angular speed of 2.0×10^{-7} rad s^{-1} at a radius of 1.5×10^{11} m.
The Earth has a mass of 6.0×10^{24} kg.

(a) Calculate the Earth's linear speed. [2 marks]

(b) (i) Calculate the centripetal force needed to keep the Earth in its orbit. [2 marks]

    (ii) What provides this centripetal force? [1 mark]

Q2 A bucket full of water, tied to a rope, is being swung around in a vertical circle (so it is upside down at the top of the swing). The radius of the circle is 1 m.

(a) At the top of the swing, all the water remains in the bucket.
What is the minimum centripetal acceleration of the bucket? [1 mark]

(b) Calculate the minimum frequency with which the bucket can be swung without any water falling out. [2 marks]

I'm spinnin' around, move out of my way...

"Centripetal" just means "centre-seeking". The centripetal force is what actually causes circular motion.
*What you **feel** when you're spinning, though, is the reaction (centrifugal) force. Don't get the two mixed up.*

Gravitational Fields

*Gravity's all about masses **attracting** each other. If the Earth didn't have a **gravitational field**, apples wouldn't fall to the ground and you'd probably be floating off into space instead of sitting here reading this page...*

Masses in a **Gravitational Field** Experience a **Force of Attraction**

1) Any object with mass will **experience an attractive force** if you put it in the **gravitational field** of another object.

2) Only objects with a **large** mass, such as stars and planets, have a significant effect. E.g. the gravitational fields of the **Moon** and the **Sun** are noticeable here on Earth — they're the main cause of our **tides**.

You can **Calculate Forces** Using **Newton's Law of Gravitation**

The **force** experienced by an object in a gravitational field is always **attractive**. It's a **vector** which depends on the **masses** involved and the **distances** between them. It's easy to work this out for **point masses** — or objects that behave as if all their mass is concentrated at the centre e.g. uniform spheres. You just put the numbers into this equation...

NEWTON'S LAW OF GRAVITATION:

$$F = -\frac{GMm}{r^2}$$

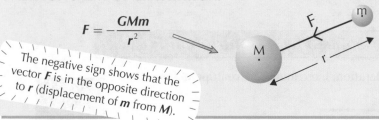

The negative sign shows that the vector **F** is in the opposite direction to **r** (displacement of **m** from **M**).

The diagram shows the force acting on **m** due to **M**. (The force on **M** due to **m** is equal but in the opposite direction.)

M and **m** behave as point masses.

G is the **gravitational constant** — 6.67×10^{-11} Nm²kg⁻².

r is the distance (in metres) between the centres of the two masses.

It doesn't matter what you call the masses: M and m, m_1 and m_2, Paul and Larry...

The law of gravitation is an **inverse square law** $\left(F \propto \frac{1}{r^2} \right)$ so:

1) if the distance **r** between the masses **increases** then the force **F** will **decrease**.

2) if the **distance doubles** then the **force** will be one **quarter** the strength of the original force.

You can Draw **Lines of Force** to Show the **Field** Around an Object

Gravitational lines of force (or "field lines") are **arrows** showing the **direction of the force** that masses would feel in a gravitational field.

1) If you put a small mass, **m**, anywhere in the Earth's gravitational field, it will always be attracted **towards** the Earth.

2) The Earth's gravitational field is **radial** — the lines of force meet at the centre of the Earth.

3) If you move mass **m** further away from the Earth — where the **lines** of force are **further apart** — the **force** it experiences **decreases**.

4) The small mass, **m**, has a gravitational field of its own. This doesn't have a noticeable effect on the Earth though, because the Earth is so much **more massive**.

5) Close to the Earth's surface, the gravitational field is (almost) uniform — so the **field lines** are (almost) **parallel**. You can usually **assume** that the field is perfectly uniform.

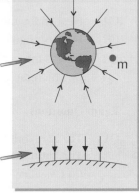

The **Field Strength** is the **Force per Unit Mass**

Gravitational field strength, **g**, is the **force per unit mass**. Its value depends on **where you are** in the field. There's a really simple equation for working it out:

 $g = \frac{F}{m}$

g has units of newtons per kilogram (Nkg⁻¹)

1) **F** is the force experienced by a mass **m** when it's placed in the gravitational field. Divide **F** by **m** and you get the **force per unit mass**.

2) **g** is a **vector** quantity, always pointing towards the centre of the mass whose field you're describing (because the gravitational force acts in that direction). This means **g** is **negative**.

3) Since the gravitational field is almost uniform at the Earth's surface, you can assume **g** is a constant.

4) **g** is just the **acceleration** of a mass in a gravitational field. It's often called the **acceleration due to gravity**.

The **value** of g at the **Earth's surface** is approximately **–9.81** ms⁻² (or –9.81 Nkg⁻¹).

Gravitational Fields

A *Mass* in a *Uniform Gravitational Field* Experiences a *Constant Force*

The **force** on an object with mass **m** in a gravitational field is given by **mg**.
In a **uniform gravitational field**, like the one near the Earth's surface, the
value of **g** is the **same** at all locations. This means that the **force** experienced
by a mass due to gravity will also be the **same** everywhere within the field.

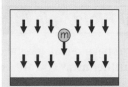

A uniform field:
field strength = **g**
force = **mg**

Potential Energy Increases with *Height* in a *Uniform Gravitational Field*

The **potential energy** of an object with mass **m** in a gravitational field is given by **mgh**, where **h** is the distance from a
certain point — e.g. the **height** above the **Earth's surface**. In a uniform field, the value of **mg** is constant whatever the
value of **h**, so:

1) **Increasing** the height of a mass, **m**, by a value of Δh changes the potential
    energy by the **same amount** wherever the mass is placed in the field.

2) You can write this relationship as an **equation**:

3) ΔPE is **positive** when the **distance increases**,
    and **negative** when the **distance decreases**.

$$\Delta PE = mg\Delta h$$

Example

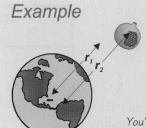

A pomegranate of mass 300 g is catapulted from a height of one metre above
the Earth's surface and reaches a height of 20 m above the Earth's surface.
The pomegranate gains potential energy as it moves away from the Earth.

Its change in potential energy is:

$g = 9.81\ ms^{-2}$

*You're only looking at the change in PE, so
you don't have to worry about the sign of **g**.*

$\Delta PE = mg\Delta h$
$= 0.3 \times 9.81 \times (20 - 1)$
$= 55.9\ J$

Practice Questions

Q1   Write down Newton's law of gravitation.

Q2   Draw a diagram showing the Earth's gravitational field:
      a) extending far from the Earth's surface, and b) close to the Earth's surface.

Q3   What is the force on a mass **m** in a gravitational field of strength **g**?

Q4   What is the equation for a change in potential energy?

Exam Questions

(Use $G = 6.67 \times 10^{-11}\ Nm^2kg^{-2}$, $g = -9.81\ Nkg^{-1}$ at the Earth's surface)

Q1   The Earth's radius is approximately 6400 km.
      Show that $M = -gr^2 / G$. Use this equation to estimate the Earth's mass.   [4 marks]

Q2   The Moon has a mass of 7.35×10^{22} kg and a radius of 1740 km.

      (a) Use Newton's law of gravitation to calculate the force on a 25 kg mass at the Moon's surface.   [2 marks]

      (b) On the Moon, $g = -1.64\ Nkg^{-1}$. A 25 kg mass is lifted 2 km above the Moon's surface.
          Calculate the gain in gravitational potential energy of the mass.   [2 marks]

If you're really stuck, put 'Inverse Square Law'...

*Clever chap, Newton, but famously tetchy. He got into fights with other physicists, mainly over planetary motion and
calculus... the usual playground squabbles. Then he spent the rest of his life trying to turn scrap metal into gold. Weird.*

Gravitational Fields

Gravity's a tricky little thing, don't you think? You know there must be a force pulling you to Earth (or whatever planet you're on), but you can't see it or feel it... or maybe you can, you just don't realise it because you've always felt it, hmm.

In a **Radial Field**, *g* is **Inversely Proportional** to *r²*

Point-like masses have **radial** gravitational fields (see the diagram on page 26). The strength of the gravitational field, *g*, **decreases** the **further away** you are from the centre of the mass, as shown on the graph below.

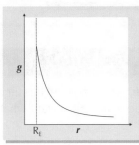

The graph shows how *g* varies for the Earth. *r* is the distance from the centre of the Earth, and R_E is the Earth's radius

The **area** under this curve gives you gravitational potential, *V* — see below.

Murray loved showing off his field strength.

You can see from the graph that the relationship between *g* and *r* isn't linear. In fact it's another **inverse square law** — make sure you know how to use it.

$$g = -\frac{GM}{r^2}$$

You gain **Gravitational Potential Energy** if you **Move Away from the Earth**

The **gravitational potential energy** of a mass, E_{grav}, is the **work** that would need to be done to move it against the force of gravity. You can express the gravitational potential energy of a mass *m* in terms of its distance *r* from a large point mass *M* by combining the equation $E_{grav} = mg\Delta h$ with the one for *g*, above.

$$E_{grav} = -\frac{GMm}{r}$$

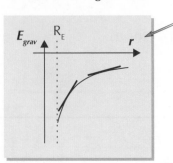

The graph shows how gravitational potential energy varies for the Earth.

1) A mass on the Earth's surface has **negative** gravitational potential energy.

2) As you move a mass away from the Earth, it **gains potential energy**.

3) Potential energy is **zero** at an **infinite** distance from the Earth.

4) The gradient of a **tangent** to the graph gives the value of the gravitational **force** at that point — the force is **greatest** at the **Earth's surface** where the graph is **steepest**.

Gravitational Potential is Potential Energy per Unit Mass

The **gravitational potential** at a point, *V*, is the **potential energy per unit mass**, $V = \frac{E_{grav}}{m}$.

In a **radial field**, the equation is: $V = -\frac{GM}{r}$

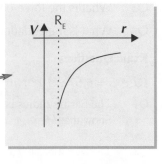

The graph of **gravitational potential** against distance has the **same shape** as the graph of **gravitational potential energy** against distance. Gravitational potential, *V*, increases with distance from the mass, and is **zero** at **infinity**.
The **gradient of a tangent** to the graph gives the value of **g** at that point.

Equipotentials show **all the points** in a field which have the **same potential**.

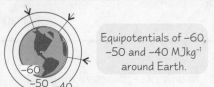

Equipotentials of –60, –50 and –40 MJkg⁻¹ around Earth.

1) If you travel along a line of equipotential you **don't lose or gain energy**.

2) For a uniform spherical mass (you can usually assume the Earth's one) the equipotentials are spherical surfaces.

3) **Equipotentials** and **field lines** are **perpendicular**.

4) At the Earth's surface, V = 63 MJkg⁻¹.

Gravitational Fields

*Planets go round and round in circles. Well, **ellipses** really, but I won't tell if you don't... anyway, it's all down to gravity.*

Planets are Satellites that Orbit the Sun

1) A **satellite** is just any **smaller mass** that **orbits** a **much larger mass**.

2) For example, the **Moon** is a satellite of the Earth, the **planets** are satellites of the Sun and man-made satellites orbit the Earth broadcasting TV signals.

3) In our Solar System, the planets have **nearly circular orbits**, so you can use the **equations of circular motion** — go back and have a look at pages 24-25 if you've forgotten them already.

The Speed of an Orbit depends on its Radius and the Mass of the Larger Body

1) Earth feels a force due to the gravitational 'pull' of the **Sun**. This force is given by Newton's law of gravitation:

(see p. 26)

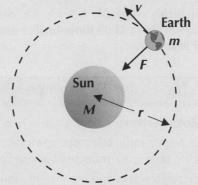

2) The Earth has velocity **v**. Its linear speed is constant but its **direction** is not — so it's accelerating. The **centripetal force** causing this acceleration is:

$$F = \frac{mv^2}{r}$$

3) The **centripetal force** on the Earth must be a result of the **gravitational force** due to the Sun, and so these forces must be **equal**...

$$\frac{mv^2}{r} = \frac{GMm}{r^2}$$ and rearranging... $$v = \sqrt{\frac{GM}{r}}$$

Practice Questions

Q1 Sketch a graph to show how the value of **g** varies with distance from the Earth's surface.

Q2 What does the area under the graph you drew in Q1 represent?

Q3 What is the difference between gravitational potential and gravitational potential energy?

Q4 What are equipotentials? What shape are the Earth's equipotentials?

Exam Questions

(Use G = 6.67 × 10⁻¹¹ Nm²kg⁻², mass of Earth = 5.98 × 10²⁴ kg, radius of Earth = 6400 km)

Q1 A satellite with a mass of 3015 kg orbits 200 km above the Earth's surface.

(a) Calculate the gravitational field strength at this distance from Earth. [2 marks]

(b) Calculate the satellite's gravitational potential energy. [2 marks]

(c) Find the linear speed of the satellite. [2 marks]

Q2 The Sun has a mass of 2.0 × 10³⁰ kg, but loses mass at a rate of around 6 × 10⁹ kgs⁻¹.
Discuss whether this will have had any significant effect on the Earth's orbit over the past 50 000 years. [2 marks]

Increase your potential – stand on a chair...

It was a charming fellow called Johannes Kepler who showed that the planets orbited in ellipses. He's also been proclaimed as the first science fiction writer — must've been a busy chap. He wrote a tale about a fantastic trip to the Moon, where the book narrator's mum asks a demon the secret of space travel, to boldly go where — oh wait, different story.

The Solar System & Astronomical Distances

The meaning of life, Part 6: A2 Physics...

Our **Solar System** Contains the **Sun**, **Planets**, **Satellites**, **Asteroids** and **Comets**

1) Our **Solar System** consists of the **Sun** and all of the objects that **orbit** it:

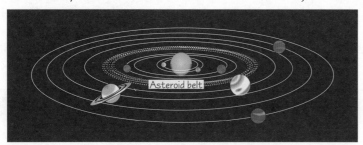

Asteroid belt

The planets (in order): **Mercury**, **Venus**, **Earth**, **Mars**, **Jupiter**, **Saturn**, **Uranus** and **Neptune** (as well as the asteroid belt) all have nearly **circular** orbits. We used to call Pluto a planet too, but it's been reclassified now.

Remember — planets, moons and comets don't emit light; they just reflect it.

2) The orbits of the **comets** we see are **highly elliptical**. Comets are "**dirty snowballs**" (lumps of rock, ice and dust) that we think usually orbit the sun about **1000 times further away** than **Pluto** does. The most famous is **Halley's comet**, which orbits in **76 years**.

Distances and **Velocities** in the Solar System can be Measured using **Radar**

1) **Distances** between objects in the **Solar System** are enormous — one way to measure them is using **radar**.

2) A **short pulse** of **radio waves** is sent from a **radio telescope** towards a distant object, e.g. planet or asteroid. When the pulse hits the surface of the object, it's **reflected** back to Earth.

3) The telescope picks up the reflected radio waves and records the **time**, *t*, taken for them to return.

4) Radio waves in space, like all electromagnetic waves, travel at the **speed of light**, *c*, so you can work out the **distance**, *d*, to the object using a variation of the formula speed = distance ÷ time:

$$2d = ct$$

It's 2d, not just d, because the pulse travels twice the distance to the object — there and back again.

5) You can also use this method to find the **average speed** of an object **relative** to Earth. You send **two pulses** separated by a certain **time interval**, to give two separate measurements of the object's **distance**. The difference between the distances shows **how far** the object has moved in the time interval — and **distance ÷ time = speed**.

6) Like most things in physics, this method is based on a couple of **assumptions**:

> 1) The **speed** of the **radio waves** is the **same** on the way **to the object** and the **way back** to the telescope.
>
> 2) The **time** taken for the **radio waves** to **reach the object** is the **same** as the **time** taken to **return**.

For these to be true, the **speed of light** must be **constant**, even though the observer and object are both **moving**, and the **object's speed** must be **much less** than the **speed of light**, so that there are no **relativistic effects** (p. 33).

7) More accurate measurements of the speed of distant objects can be made using **Doppler shifts** (see p. 32).

You Can Use the **Brightness of Stars** to **Calculate Distances**

1) The **brightness** of a star in the night sky depends on **two** things — its **luminosity** (i.e. how much **light energy** it gives out in a given time) and its **distance from us** (if you ignore weather, light pollution, etc.). So the **brightest** stars will be **close** to us and have a **high luminosity**.

2) How **bright** a star looks when seen from **Earth** is called its **apparent magnitude** — this depends on its **absolute magnitude** (i.e. how bright it really is) and how far away it is.

3) So, to find the **distance** to a star, you need to measure how bright it looks (**apparent magnitude**) and calculate how bright it really is (**absolute magnitude**) — and put these into a lovely equation (that you don't need to know).

4) This method works for objects that you can calculate the brightness of **directly** — called **standard candles**. **Cepheid variable stars** are examples of standard candles because their brightness changes in a certain pattern. So, if you find a Cepheid variable within a galaxy, you can work out how far that galaxy is from us.

The Solar System & Astronomical Distances

Distances in the Solar System are Often Measured in Astronomical Units (AU)

1) From **Copernicus** onwards, astronomers were able to work out the **distance** the **planets** are from the Sun **relative** to the Earth, using **astronomical units** (AU). But they could not work out the **actual distances**.

> One **astronomical unit** (AU) is defined as the **mean distance** between the **Earth** and the **Sun**.

2) The **size** of the AU wasn't accurately known until 1769 — when it was carefully **measured** during a **transit of Venus** (when Venus passed between the Earth and the Sun).

Another Measure of Distance is the Light-Year (ly)

1) All **electromagnetic waves** travel at the **speed of light**, *c*, in a vacuum ($c = 3.00 \times 10^8$ ms⁻¹).

> The **distance** that electromagnetic waves travel through a vacuum in **one year** is called a **light-year** (ly).

2) If we see the light from a star that is, say, **10 light-years away** then we are actually seeing it as it was **10 years ago**. The further away the object is, the further **back in time** we are actually seeing it.

3) **1 ly** is equivalent to about **63 000 AU**.

Important Sizes and Conversions

Distances in space are tricky to make sense of because they're all **absolutely enormous**. Have a look at the **diagrams** below and you should get some idea of the **scale** of things.

Unit of Distance	Astronomical Unit (AU)	Light Year (ly)
Approximate Length in metres	1.50×10^{11}	9.46×10^{15}

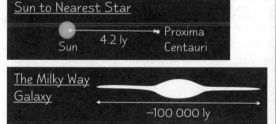

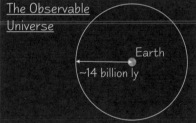

> When we look at the stars we're looking **back in time**, and we can only see as far back as the **beginning of the Universe**.
>
> So the size of the **observable Universe** is the **age** of the Universe multiplied by the **speed of light**.

Practice Questions

Q1 What is meant by a) an astronomical unit, b) a light-year?

Q2 How do we measure the distance to objects in the Solar System using radar?

Exam Questions

Q1 A pulse of radio waves is transmitted from a telescope on Earth towards Venus. The telescope detects the radio waves returning after 4.6 minutes.

    (a) Show that the distance from Earth to Venus at this time is approximately 4.1×10^{10} km   [3 marks]

    (b) State two assumptions that your calculation in part (a) is based on.   [2 marks]

Q2 (a) Give the definition of a *light-year*.   [1 mark]

    (b) Light travels at 3.0×10^8 ms⁻¹ in a vacuum. Calculate the distance of a light-year in metres.   [2 marks]

    (c) Explain why the size of the observable Universe is limited by the speed of light.   [2 marks]

So — using a ruler's out of the question then...

Don't bother trying to get your head round these distances — they're just too big to imagine. Just learn the powers of ten and you'll be fine. Make sure you understand how to measure distances using radar though — that bit does make sense.

The Doppler Effect and Redshift

Another (and sadly more confusing) way to measure the speed of distant objects uses the Doppler effect.

You Can **Measure** the **Speed** of Objects Using the **Doppler Effect**

1) If you stand **still** and listen to the sound of the horn of a **stationary** car, you'll hear the **same pitch** sound no matter where you stand.

2) But if the car is **moving** when it sounds its horn, the pitch you hear will be **different** — it'll be **lower** if the car is moving **away** from you and **higher** if it's moving **towards** you. This is the **Doppler effect**.

3) When the car is moving **away** from you, the sound waves travel in the **opposite** direction from the car, so are **stretched** out — i.e. have a **longer wavelength** and **lower frequency** when they reach you.

4) The opposite happens when the car is moving **towards** you — the sound waves **bunch up**, so have a **shorter wavelength** and **higher frequency** when they get to you.

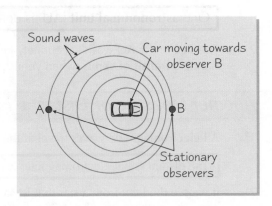

"That star's going well over 70 mph — let's book him."

5) How much the sound waves change depends on how **fast** the car is travelling — the **greater** the car's **speed**, the **larger** the **change**. This means the Doppler effect can be used to **measure the speed** of moving objects.

6) The Doppler effect happens with **all waves**, so it has a wide range of applications. For example, **police radar guns** measure the speed of cars using **microwaves**, and **cosmologists** measure the speed of distant **stars** using the **light** they **emit**.

The **Doppler Effect** Means **Radiation Emitted** by **Distant Objects** is **"Shifted"**

You can calculate the **speed** of distant objects relative to the Earth by measuring how their **movement** affects the **radiation** they **emit**. The method is based on the principle that an **atom** will emit and absorb radiation with the **same**, **characteristic spectrum** (page 33) wherever it is — i.e. on **Earth** or in a **distant star**.

For example, if you detect the **radiation** emitted by a **hydrogen atom** in a **star** and compare it with the radiation emitted by a **hydrogen atom** on **Earth**, you can calculate how much the radiation is **"shifted"** and so determine the **velocity** of the star relative to the Earth using the **equation** given below.

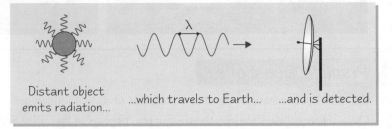

Distant object emits radiation...   ...which travels to Earth...   ...and is detected.

How the **radiation** is **shifted** depends on the movement of the object:

1) When an object is moving **away** from Earth, the wavelengths of its radiation get **longer** and the frequencies get lower — i.e. it shifts towards the **red** end of the spectrum, so this is **redshift**.

2) When an object is moving **towards** Earth, the **opposite** happens and the radiation undergoes **blueshift**.

The **equation** for finding the velocity of an object from the radiation it emits and absorbs is:

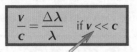

$$\frac{v}{c} = \frac{\Delta\lambda}{\lambda} \quad \text{if } v \ll c$$

$v \ll c$ means "v is much less than c" (see page 33 for why this is important).

where $\Delta\lambda$ is the change in the wavelength, λ is the wavelength of the emitted radiation, v is the velocity of the object and c is the speed of light.

The Doppler Effect and Redshift

Stars Produce Absorption Spectra

1) Radiation is emitted from a **very hot** region of a star (called the **photosphere**) in a **continuous spectrum**.

2) **Atoms** in the atmosphere of the star **absorb** certain **wavelengths** of the radiation, producing dark **absorption lines** within the spectrum.

3) Different atoms absorb different parts of the spectrum, resulting in a **characteristic pattern** for each atom. By looking at the absorption lines from a star, the **composition** of the **stellar atmosphere** can be worked out.

   The **spectrum** from the **star** is compared with **known spectra** in the lab: ⟹

   Stellar spectrum (containing H, He and Na)
   Hydrogen
   Helium
   Sodium

4) Once you've worked out which **atoms** make up the **pattern of absorption lines**, you can compare the **position** of the **absorption lines** for **each atom** in the **star's spectrum** with the **same spectrum** recorded in the **lab**.

   This shows **how much** the **spectrum** has been **shifted** by the movement of the star.

   For example, the spectrum from this star is ⟶ **shifted** towards the **red end** of the spectrum, showing that it is **moving away** from Earth.

   Stellar spectrum
   Lab spectrum

Time Dilation Happens Close to the Speed of Light

1) A big **assumption** of the methods on the previous few pages is that the **speed** of the object being studied is **much less** than the **speed of light**, **c**.

2) This matters because **time** runs at **different speeds** for two objects **moving relative** to each other — but it's only really noticeable **close to** the **speed of light** (which is why you can **ignore** it as long as the object isn't travelling too fast).

   It's not just time that starts doing weird things at high speeds, mass and energy do too (see page 67).

3) This effect is called **relativistic time dilation**.

   A **stationary** observer measures the time interval between two events as t_0, the **proper time**. An observer moving at a **constant velocity**, v, will measure a **longer** interval, t, between the two events. t is given by the equation:

   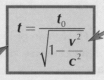
   $$t = \frac{t_0}{\sqrt{1 - \frac{v^2}{c^2}}}$$

   $\frac{1}{\sqrt{1 - \frac{v^2}{c^2}}}$ is called the **relativistic factor**, γ.

   When $v \ll c$, $\gamma \approx 1$, so the time measured by both observers is almost the same, $t = t_0$ (which is why you can ignore relativistic effects at low speeds).

Practice Questions

Q1 What is the Doppler effect?

Q2 What is the formula for working out the speed of a distant object from the radiation it emits?

Q3 What does '$v \ll c$' mean?

Q4 What is relativistic time dilation?

Exam Question

Q1 The radiation emitted by objects A and B has been detected.

   (a) The light from object A has been shifted towards the blue end of the spectrum. What does this suggest about object A's movement? [1 mark]

   (b) The wavelength of a hydrogen emission in object B's spectrum is 667.83 nm. In the laboratory, the wavelength of the same emission is measured as 656.28 nm. Calculate object B's velocity relative to the Earth and state its direction of travel. [3 marks]

Time dilation occurs close to the speed of light — and in history lessons...

Ok... so time dilation isn't the easiest thing to get your head around, but I'm afraid you need to know it folks. Luckily it only happens close to the speed of light, and that as long as the assumption '$v \ll c$' is true it's pretty safe to ignore it.

The Big Bang Model of the Universe

Right, we're moving on to the BIG picture now — we all like a bit of cosmology...

Hubble Realised that Recessional Velocity is Proportional to Distance

1) The **spectra** from **galaxies** (apart from a few very close ones) all show **redshift** (see page 32).
The amount of **redshift** gives the **recessional velocity** — how fast the galaxy is moving away from Earth.

2) When **Hubble** plotted the **recessional velocity** of galaxies against their **distance** from Earth (found using Cepheid variables — see page 30) he found that they were **proportional**. This gave rise to **Hubble's law**:

$$v = H_0 d$$

where v = recessional velocity in **kms^{-1}**, d = distance in **Mpc** and H_0 = **Hubble's constant** in **kms^{-1}Mpc^{-1}**.

A megaparsec (Mpc) is just a unit of distance used in astronomy.
1 Mpc = 3.09 × 10^{22} m

3) Since distance is very difficult to measure, astronomers disagree on the value of H_0. It's generally accepted that H_0 lies somewhere between 50 kms^{-1}Mpc^{-1} and 100 kms^{-1}Mpc^{-1}.

4) The **SI unit** for H_0 is s^{-1}. To get H_0 in SI units, you need v in ms^{-1} and d in m (1 Mpc = 3.09 × 10^{22} m).

The Universe is Expanding

1) By showing that objects in the Universe are **moving away** from each other, Hubble's work is strong evidence that the **Universe is expanding**. The rate of expansion depends on the value of H_0, Hubble's constant.

2) The way cosmologists tend to look at this, the galaxies aren't actually moving **through space** away from us. Instead, **space itself** is expanding and the light waves are being **stretched** along with it. This is called **cosmological redshift** to distinguish it from **redshift** produced by sources that **are** moving through space. Don't worry though — you can use the **formula** on page 32 for **both kinds** of redshift.

3) Since the Universe is **expanding uniformly** away from **us** it seems as though we're at the **centre** of the Universe, but this is an **illusion**. You would observe the **same thing** at **any point** in the Universe.

The Age and Observable Size of the Universe Depend on H₀

1) If the Universe has been **expanding** at the **same rate** for its whole life, the **age** of the Universe is $t = 1/H_0$ (time = distance/speed). This is only an estimate since the Universe probably hasn't always been expanding at the same rate.

2) Since no one knows the **exact value** of H_0 we can only guess the Universe's age. If H_0 = **75 kms^{-1}Mpc^{-1}**, then the age of the Universe ≈ 1/(2.4 × 10^{-18} s^{-1}) = 4.1 × 10^{17} s, which is approximately **13 billion years**.

3) The **absolute size** of the Universe is **unknown** (and changing) but there is a limit on the size of the **observable Universe**. This is simply a **sphere** (with the Earth at its centre) with a **radius** equal to the **maximum distance** that **light** can travel during its **age**. So if H_0 = **75 kms^{-1}Mpc^{-1}** then this sphere will have a radius of **13 billion light years**.

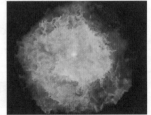

NB 13 billion candles is too many to put on a birthday cake.

The Redshift of Galaxies is Strong Evidence for the HBB

1) If the Universe is **expanding**, then further back in time it must have been much **smaller**. In fact, if you trace time back **far enough**, and assume that the **expansion** has always been happening, then the entire Universe must once have been contained in a **single point** — this is the basis of the **Hot Big Bang theory**.

> **THE HOT BIG BANG THEORY (HBB):** the Universe started off **very hot** and **very dense** (perhaps as an **infinitely hot, infinitely dense** singularity) and has been **expanding** ever since.

2) The HBB theory is currently the **best explanation** we've got for the state of the Universe and is widely **accepted**.

3) Since Hubble published his findings, the **redshifts** of other, more distant objects in the **Universe** have been measured — and they **continue** to fit the **pattern** he predicted, providing more **evidence** for the **HBB theory**. And it doesn't stop there — there's even more evidence for the HBB theory on the next page.

The Big Bang Model of the Universe

Cosmic Microwave Background Radiation — More Evidence for the HBB

1) The Hot Big Bang model predicts that loads of **electromagnetic radiation** was produced in the **very early Universe**. This radiation should **still** be observed today (it hasn't had anywhere else to go).

2) Because the Universe has **expanded**, the wavelengths of this cosmic background radiation have been **stretched** and are now in the **microwave** region.

3) This was picked up **accidentally** by Penzias and Wilson in the 1960s.

Properties of the Cosmic Microwave Background Radiation (CMBR)

1) In the late 1980s a satellite called the **Cosmic Background Explorer** (**COBE**) was sent up to have a **detailed look** at the radiation.

2) It found a **continuous spectrum** corresponding to a **temperature** of **2.73 K**.

3) The radiation is largely **isotropic** and **homogeneous** — it's about the **same intensity** whichever direction you look.

4) There are **very tiny fluctuations** in temperature, which were at the limit of COBE's detection. These are due to tiny energy-density variations in the early Universe, and are needed for the initial '**seeding**' of galaxy formation.

5) The background radiation also shows a **Doppler shift**, indicating the Earth's motion through space. It turns out that the **Milky Way** is rushing towards an unknown mass (the **Great Attractor**) at over a **million miles an hour**.

Another Bit of Evidence is the Amount of Helium in the Universe

1) The HBB model also explained the **large abundance of helium** in the Universe (which had puzzled astrophysicists for a while).

2) The early Universe had been very hot, and at some point it must have been hot enough for **hydrogen fusion** to happen. This means that, together with the theory of the synthesis of the **heavier elements** in stars, the **relative abundances** of all of the elements can be accounted for.

Practice Questions

Q1 What is cosmological redshift? How is it different from ordinary redshift?

Q2 Outline the hot big bang theory.

Q3 What is the cosmic background radiation?

Exam Questions

Q1 (a) What does Hubble's law suggest about the nature of the Universe? [2 marks]

(b) Assume $H_0 = +50$ kms^{-1}Mpc^{-1} (1 Mpc = 3.09×10^{22} m).

   i) Calculate H_0 in SI units. [2 marks]

   ii) The age of the Universe is approximately equal to H_0^{-1}. Use your value from (b) i) to estimate the age of the Universe, and hence the size of the observable universe. [3 marks]

Q2 (a) A galaxy has a redshift of 0.37. Estimate the speed at which it is moving away from us. [2 marks]

(b) Use Hubble's law to estimate the distance (in light years) that the galaxy is from us.
(Take $H_0 = 2.4 \times 10^{-18}$ s^{-1}, 1 ly = 9.5×10^{15} m.) [2 marks]

(c) Explain why the speed of the galaxy means that your answers to (a) and (b) are only estimates. [1 mark]

Q3 The cosmic microwave background is a continuous spectrum at a temperature of about 3 K.
Explain why its discovery was considered strong evidence for the Hot Big Bang model of the Universe. [2 marks]

My Physics teacher was a Great Attractor — everyone fell for him...

The simple Big Bang model doesn't actually work — not quite, anyway. There are loads of little things that don't quite add up. Modern cosmologists are trying to improve the model using a period of very rapid expansion called inflation.

Ideal Gases

*Aaahh... great... another one of those 'our equation doesn't work properly with **real gases**, so we'll invent an **ideal** gas that it **does work** for and they'll think we're dead clever' situations. Hmm. Physicists, eh...*

There's an **Absolute Scale** of **Temperature**

There is a **lowest possible temperature** called **absolute zero***. Absolute zero is given a value of **zero kelvin**, written **0 K**, on the absolute temperature scale.

At **0 K** all particles have the **minimum** possible **kinetic energy** — everything pretty much stops — at higher temperatures, particles have more energy. In fact, with the **Kelvin scale**, a particle's **energy** is **proportional** to its **temperature** (see page 41).

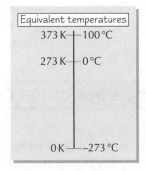

Equivalent temperatures

373 K —— 100 °C

273 K —— 0 °C

0 K —— −273 °C

1) The Kelvin scale is named after Lord Kelvin who first suggested it.

2) A change of **1 K** equals a change of **1 °C**.

3) To change from degrees Celsius into kelvin you **add 273** (or 273.15 if you need to be really precise).

$$K = C + 273$$

It's true. −273.15 °C is the lowest temperature theoretically possible. Weird, huh. You'd kinda think there wouldn't be a minimum, but there is.

All **equations** in **thermal physics** use temperatures measured in kelvin.

There are **Three Gas Laws**

The three gas laws were each worked out **independently** by **careful experiment**. Each of the gas laws applies to a **fixed mass** of gas.

Boyle's Law

At a **constant temperature** the **pressure *p*** and **volume *V*** of a gas are **inversely proportional**.

A (theoretical) gas that obeys Boyle's law at all temperatures is called an **ideal gas**.

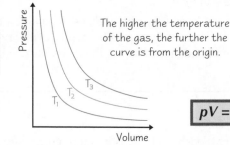

The higher the temperature of the gas, the further the curve is from the origin.

$$pV = \text{constant}$$

Charles' Law

At constant **pressure**, the **volume *V*** of a gas is **directly proportional** to its **absolute temperature *T***.

Ideal gases obey this law and the pressure law as well.

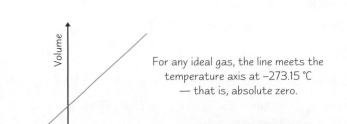

For any ideal gas, the line meets the temperature axis at −273.15 °C — that is, absolute zero.

$$V/T = \text{constant}$$

'Ello, 'ello...

If you'd plotted these graphs in kelvin, they'd both have gone through the origin.

The Pressure Law

At constant **volume**, the **pressure *p*** of a gas is **directly proportional** to its **absolute temperature *T***.

$$p/T = \text{constant}$$

Ideal Gases

If you Combine All Three you get the Ideal Gas Equation

Combining all three gas laws gives the equation: $\dfrac{pV}{T} = $ constant

1) The constant in the equation depends on the amount of gas used. ← *(Pretty obvious... if you have more gas it takes up more space.)*
   The amount of **gas** can be **measured** in **moles**, *n*.

2) The constant then becomes *nR*, where *R* is called the **molar gas constant**.
   Its value is 8.31 J mol⁻¹ K⁻¹.

3) Plugging this into the equation gives: $\dfrac{pV}{T} = nR$ or rearranging, *pV = nRT — the ideal gas equation*

This equation works well (i.e., a real gas approximates to an ideal gas)
for gases at **low pressure** and fairly **high temperatures**.

Boltzmann's Constant k is like a Gas Constant for One Particle of Gas

One mole of any **gas** contains the same number of particles.
This number is called **Avogadro's constant** and has the symbol N_A. The value of N_A is **6.02 × 10²³ particles per mole**.

1) The **number of particles** in a **mass of gas** is given by the **number of moles**, *n*, multiplied by **Avogadro's constant**.
   So the number of particles, $N = nN_A$.

2) **Boltzmann's constant**, *k*, is equivalent to R/N_A — you can think of Boltzmann's constant as the **gas constant** for **one particle of gas**, while *R* is the gas constant for **one mole of gas**.

3) The value of Boltzmann's constant is **1.38 × 10⁻²³ JK⁻¹**.

4) If you combine $N = nN_A$ and $k = R/N_A$ you'll see that $Nk = nR$
   — which can be substituted into the ideal gas equation: → *pV = NkT — the equation of state*

   The equation *pV = NkT* is called the equation of state of an ideal gas.

Practice Questions

Q1 State Boyle's law, Charles' law and the pressure law.

Q2 What is the ideal gas equation?

Q3 The pressure of a gas is 100 000 Pa and its temperature is 27 °C. The gas is heated — its volume
   stays fixed but the pressure rises to 150 000 Pa. Show that its new temperature is 177 °C.

Q4 What is the equation of state of an ideal gas?

Exam Questions

Q1 At a constant volume, the pressure and absolute temperature of a gas will be proportional.
   Describe two features of a graph of pressure (Pa) against temperature (K) that show this relation. [1 mark]

Q2 The mass of one mole of nitrogen gas is 0.028 kg. R = 8.31 J mol⁻¹ K⁻¹.

   (a) A flask contains 0.014 kg of nitrogen gas. How many moles of nitrogen gas are in the flask? [1 mark]

   (b) The flask has a volume of 0.01 m³ and is at a temperature of 27 °C. Calculate the pressure inside it. [2 marks]

   (c) Describe how the pressure inside the flask would change if the number of molecules inside was halved. [1 mark]

Q3 A large helium balloon has a volume of 10 m³ at ground level.
   The temperature of the gas in the balloon is 293 K and the pressure is 1 × 10⁵ Pa.
   The balloon is released and rises to a height where its volume becomes 25 m³ and its temperature is 260 K.
   Calculate the pressure inside the balloon at its new height. [3 marks]

Ideal revision equation — marks = (pages read × questions answered)²...

*All this might sound a bit theoretical, but most gases you'll meet in the everyday world come fairly close to being 'ideal'.
They only stop obeying these laws when the pressure's too high or they're getting close to their boiling point.*

The Pressure of an Ideal Gas

Kinetic theory tries to **explain** the **gas laws**. It basically models a gas as a series of hard balls that obey Newton's laws. Luckily for you, you don't have to derive the ideal gas equation — but you do have to use it...

Imagine *a Particle in a Box*...

To understand what's going on in the ideal gas equation, it can really help to think about gas particles in a box (trust me...)

Imagine a cubic box with sides of length **l** containing one particle, Q, of mass **m**.

Say particle Q **moves horizontally** towards **wall A** with velocity **u**. Its **momentum** approaching the wall is ***mu***. It strikes wall **A**, exerting a **force** on the wall, and heads back in the opposite direction.

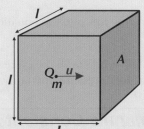

1) ### Particle velocity is proportional to the pressure

The **faster** the particle, the **larger** its **momentum**, so the **larger** the **force** on the wall. The **particle** will also take **less time** to travel across the box and back again, and so will hit the walls more often. And as **pressure = force ÷ area**, the **pressure** will be **greater** too.

2) ### The number of particles, N, is proportional to the pressure

Instead of just one particle, imagine you've got a whole stream of them hitting wall A. Each particle exerts a force on the wall as it hits it, so the total force on the wall will be proportional to the number of particles. And you've guessed it, as pressure = force ÷ area, the pressure is proportional too.

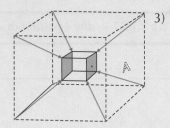

3) ### The volume of the box is inversely proportional to the pressure

Now imagine you shrink the box. The particles have **less distance** to travel before they hit a wall, so you've **increased** the number of times the particles hit the walls of the box per second, which increases the total force on the wall.

Because the box is now smaller, the **area** of the walls is **smaller**. So there's a greater force on a smaller area, meaning the **pressure is greater**.

4) ### Particles travel in random directions at different velocities

Obviously gas particles don't all neatly go in the same direction — you can estimate that a third of all the particles are travelling in one dimension (x, y, z) at any time (as there are three dimensions).

To take account of the different particle velocities, you take the average of the squares of the velocities, called the mean square speed, $\overline{c^2}$.

If you do the maths, it all whittles down to give you this amazing **equation**:

$$pV = \frac{1}{3}Nm\overline{c^2}$$

A Useful Quantity is the *Root Mean Square Speed* $\sqrt{\overline{c^2}}$

$\overline{c^2}$ is the **mean square speed** and has **units m²s⁻²**. It helps to think about the motion of a typical particle.

1) $\overline{c^2}$ is the **square** of the **speed** of an **average particle**, so the square root of it gives you the typical speed.

2) This is called the **root mean square speed** or, usually, the **r.m.s. speed**.

$$r.m.s.\ speed = \sqrt{mean\ square\ speed} = \sqrt{\overline{c^2}}$$

The Pressure of an Ideal Gas

Lots of **Simplifying Assumptions** are Used in **Kinetic Theory**

In **kinetic theory**, physicists picture gas particles moving at **high speed** in **random directions**.
To get **equations** like the one on the previous page, some **simplifying assumptions** are needed:

1) The gas contains a **large number of particles**.
2) The particles **move rapidly** and **randomly**.
3) The motion of the particles follows **Newton's laws**.
4) **Collisions** between particles themselves or at the walls of a container are **perfectly elastic**.
5) There are **no attractive forces** between particles.
6) Any **forces** that act during collisions are **instantaneous**.
7) Particles have a **negligible volume** compared with the volume of the container.

A **gas obeying** these **assumptions** is called an **ideal** gas. Real gases behave like ideal gases as long as
the **pressure isn't too big** and the **temperature** is **reasonably high** (compared with their boiling points).

Each Particle in a Gas Goes on a **Random Walk**

1) There's no way you can **record** the **random motion** of all the particles in a **gas** (without going cross-eyed in the process, of course). Instead you can **model** the movement of the particles by a **random walk**.

2) A **random walk** assumes that each **particle** starts in one place, **moves N steps** in random directions, and ends up **somewhere else**.

3) Here's the path taken by one particle in a box filled with air. The particle **changes direction** each time it **collides** with another particle in the box.

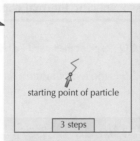
starting point of particle
3 steps

10 steps

100 steps

4) What's really useful is that the average **distance moved** in those N steps is proportional to $\sqrt{N}$.

5) The distance a particle can travel between collisions is usually around 10^{-7} m .
So to travel 1 m from its starting point, a particle will have had to take 10 000 000 steps. That's quite a few... so it's no wonder that diffusion is a **slow process**, even if the particles are travelling at high speeds.

Practice Questions

Q1 What is the definition of the mean square speed for N particles?
Q2 State the seven assumptions made about ideal gas behaviour.
Q3 What is meant by the random walk of a gas particle?
Q4 A smoke particle takes a random walk of 10 000 steps. How far has it travelled from its starting point?

Exam Question

Q1 Some helium gas is contained in a flask of volume 7×10^{-5} m³. Each helium atom has a mass of 6.6×10^{-27} kg, and there are 2×10^{22} atoms present. The pressure of the gas is 1×10^5 Pa.

    (a) Use the equation $pV = \frac{1}{3}Nm\overline{c^2}$ to show that the mean square speed of the atoms is 159 091 ms⁻¹. [2 marks]

    (b) Calculate the r.m.s. speed of the atoms. [1 mark]

    (c) The absolute temperature of the gas is doubled. Use ideas about particle velocity and momentum to explain why the pressure of the gas will have increased. [3 marks]

Doin' the random walk.... Oi!

Mean square speed is the average (mean) of the squared speeds. To find its value square all the speeds and then find the average. Don't make the mistake of finding the average speed first and then squaring. Cos that would be, like, soooo stupid.

Internal Energy and Temperature

*The energy of a particle depends on its temperature on the **thermodynamic scale** (that's Kelvin to you and me).*

If **A** and **B** are in **Thermal Equilibrium** with **C, A** is in **Equilibrium** with **B**

If **body A** and **body B** are both in **thermal equilibrium** with **body C**, then **body A** and **body B** must be in thermal equilibrium with **each other**.

This is linked with the idea of **temperature**.

1) Suppose A, B and C are three identical metal blocks. A has been in a **warm oven**, B has come from a **refrigerator** and C is at **room temperature**.

2) **Thermal energy** flows from A to C and C to B until they all reach **thermal equilibrium** and the net flow of energy stops. This happens when the three blocks are at the **same temperature**.

Thermal energy is **always** transferred from regions of **higher temperature** to regions of **lower temperature**.

*Specific Thermal Capacity is how much **Energy** it Takes to **Heat** Something*

When you heat something, its particles get more **kinetic energy** and its **temperature** rises.

The **specific thermal capacity** (*c*) of a substance is the amount of **energy** needed to **raise** the **temperature** of **1 kg** of the substance by **1 K** (or 1°C).

or put another way: **energy change = mass × specific thermal capacity × change in temperature**

in symbols: $\Delta E = mc\Delta\theta$ ←— ΔQ is sometimes used instead of ΔE for the change in thermal energy.

ΔE is the energy change in J, *m* is the mass in kg and $\Delta\theta$ is the temperature change in K or °C. Units of **c** are $J\,kg^{-1}\,K^{-1}$ or $J\,kg^{-1}\,°C^{-1}$.

*The **Speed Distribution** of **Gas Particles** Depends on **Temperature***

The **particles** in a **gas don't** all **travel** at the **same speed**. Some particles will be moving fast but others much more slowly. Most will travel around the average speed. The shape of the **speed distribution** depends on the **temperature** of the gas.

As the temperature of the gas increases:

1) the **average** particle speed increases.

2) the **maximum** particle speed increases.

3) the distribution curve becomes more **spread out**.

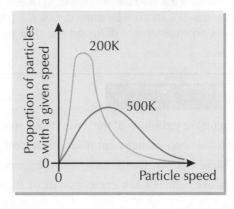

Energy Changes Happen Between Particles

The particles of a gas **collide** with each other **all the time**. Some of these collisions will be 'head-on' (particles moving in **opposite directions**) while others will be '**shunts from behind**' (particles moving in the **same direction**).

1) As a result of the collisions, **energy** will be **transferred** between particles.

2) Some particles will **gain speed** in a collision and others will **slow down**.

3) **Between collisions**, the particles will travel at **constant speed**.

4) Although the energy of an individual particle changes at each collision, the collisions **don't alter** the **total energy** of the **system**.

5) So, the **average** speed of the particles will stay the same provided the **temperature** of the gas **stays the same**.

Internal Energy and Temperature

Internal Energy is the Sum of Kinetic and Potential Energy

All things (solids, liquids, gases) have **energy** contained within them. The amount of **energy** contained in a system is called its **internal energy** — it's found by **summing** the **kinetic** and **potential energy** of all the **particles** within it.

> **Internal energy** is the **sum** of the **kinetic** and **potential energy** of the **particles** within a system.

For example, the **internal energy** of an **ideal gas** is due to the **kinetic energy** of the **particles** within it. But, how do you **sum** the **individual energies** when the particles all move at **different speeds**, so have **different kinetic energies**? The answer is to find the **average kinetic energy** of a particle, then **multiply** by the number of particles.

Average Kinetic Energy is Proportional to Absolute Temperature

There are **two equations** for the **product pV** of a gas — the ideal gas equation (page 37), and the equation involving the mean square speed of the particles (page 38). You can **equate these** to get an expression for the **average kinetic energy**.

1) $\frac{1}{2}m\overline{c^2}$ is the **average kinetic energy** of an **individual particle**.

2) The **internal energy** of an ideal gas is the **product** of the **average kinetic energy** of its particles and the **number of particles** within it. So the **average kinetic energy** is directly proportional to the **absolute temperature**.

$$\frac{1}{2}m\overline{c^2} = \frac{3}{2}\frac{nRT}{N}$$

c is the velocity of a particle. $\overline{c^2}$ is the average of the squared speeds of the particles, called the <u>mean square speed</u>.

3) So the **internal energy** must also be **dependent** on **temperature**.

> A **rise** in the **absolute temperature** will cause an **increase** in the **kinetic energy** of each particle, meaning a rise in **internal energy**.

Practice Questions

Q1 Describe the changes in the distribution of gas particle speeds as the temperature of a gas increases.

Q2 What is internal energy? What would cause a rise in internal energy?

Q3 What happens to the average kinetic energy of a particle if the temperature of a gas doubles?

Q4 Water has a specific thermal capacity of 4184 J kg^{-1} K^{-1}.
Show that the energy needed to change the temperature of 3 kg of water from 15 °C to 17 °C is 25104 J.

Exam Questions

Q1 The mass of one mole of nitrogen molecules is 2.8×10^{-2} kg. There are 6.02×10^{23} molecules in one mole.
(a) Find the mass of one molecule. [1 mark]
(b) Show that the typical speed of a nitrogen molecule at 300 K is approximately 520 ms^{-1}. [3 marks]
(c) Explain why all the nitrogen molecules will not be moving at this speed. [2 marks]

Q2 Some air freshener is sprayed at one end of a room. The room is 8.0 m long and the temperature is 20 °C.
(a) The average freshener molecule moves at 400 ms^{-1}. How long would it take for a particle to travel directly to the other end of the room? [1 mark]
(b) The perfume from the air freshener only slowly diffuses from one end of the room to the other. Explain why this takes much longer than suggested by your answer to part (a). [2 marks]
(c) How would the speed of diffusion be different if the temperature was 30 °C? Explain your answer. [3 marks]

Positivise your internal energy, man...

Phew... there's a lot to take in on these pages. Go back over it, step by step, and make sure you understand it all: specific thermal capacity, speed distribution of gas particles, internal energy and the average kinetic energy of a gas particle.

Activation Energy

Welcome to the big bad world of statistical physics — Ludwig Boltzmann's got a lot to answer for...

The Average Thermal **Energy** of a Particle is Proportional to the **Temperature**

1) Any particle above absolute zero has some **thermal energy**.

> The **average thermal energy per particle** is (very roughly) kT.

k is <u>Boltzmann's constant</u>, $k = 1.38 \times 10^{-23}$ JK^{-1}.
T is the temperature in kelvin. See page 36 for more.

2) This table gives you an idea of the magnitude of the thermal energy at various temperatures:

Temperature (K)	average thermal energy (approx.) — kT		
	J (per particle)	J mol^{-1}	eV (per particle)
1	1×10^{-23}	8	9×10^{-5}
300 (room temp)	4×10^{-21}	2000	0.03
6000 (Sun's surface)	8×10^{-20}	5×10^4	0.5

To convert kT to J mol^{-1}, multiply by Avogadro's constant (6.02×10^{23} particles per mole).

To convert kT to eV (electron-volts), divide by the charge on the electron (1.6×10^{-19} C). (See p 68 for more on electron-volts.)

3) Particles in matter are **held together** by **bonds**. The **energy** needed to break these bonds in a given substance is the **activation energy** ε (the Greek "epsilon").

4) The ratio ε/kT is really important. When kT is **big enough** compared with ε, the bonds are broken and the matter comes apart.

Activation Energy is the Energy Needed to Make Something Happen

1) For a process like a change of state to happen, particles need to 'climb' an **energy barrier**.

2) The **activation energy**, ε, is the **energy needed** to climb that barrier.

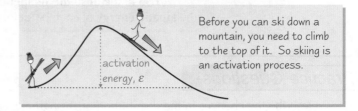

Before you can ski down a mountain, you need to climb to the top of it. So skiing is an activation process.

Lots of Processes have an **Activation Energy**

1) Lots of processes involving **particles** have activation energies — for example:

> a) **A change of state:** the particles need enough energy to break the forces between them.
>
> b) **Thermionic emission:** if you heat up a conductor, electrons are released from the surface. These electrons need enough energy to escape from the attraction of the positive nuclei.
>
> c) **Ionisation in a candle flame:** the molecules in the air need enough energy to split up into individual atoms and ions. This is a similar process to thermionic emission.
>
> d) **Conduction in a semiconductor:** semiconductors will only start to conduct once there are electrons in a high-energy state called the "conduction band", so electrons need enough energy to jump from the ground state to this higher-energy state.
>
> e) **Viscous flow:** viscous fluids have strong attractive forces between the particles, causing the fluid to 'flow' slowly. As you increase the temperature you increase the kinetic energy of the particles. This means they have more energy to overcome these forces and so the fluid will be able to flow more easily.
>
> That's why when you've got cold oil in a pan the oil is fairly viscous and doesn't 'run' very easily. As the pan and oil heat up, the oil will flow around the pan much more easily.

2) In each of these examples, the **activation energy**, ε, comes from the **random thermal energy** of the particles. You might think, then, that these processes wouldn't happen unless $kT \geq \varepsilon$... but it's not that simple...

Activation Energy

Getting **Extra Energy** is all about **Random Collisions**

1) If the **ratio** between the activation energy and the average energy of the particles (ε/kT) is too high, nothing happens.

2) As ε/kT gets down to somewhere around **15–30**, the process starts to happen at a **fair rate**.

3) So some particles must have energies of **15–30 times** the **average energy**.

4) Every time particles **collide**, there's a **chance** that one of them will gain **extra energy** — above and beyond the average **kT**. If that happens **several times** in a row, a particle can gain energies **much, much higher** than the average.

5) Say **f** is the fraction of particles with an extra energy **E**. If **E** is reasonably big compared to **kT**, then **f** will be **small**. (So far so good.)

6) Now to get a particle with an extra energy of 2**E**, you need a collision between two particles with an **average extra energy** of about **E**.

So the fraction of particles with an extra energy of 2**E** will be:

$$f \times f = f^2$$

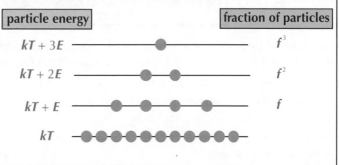

particle energy	fraction of particles
$kT + 3E$	f^3
$kT + 2E$	f^2
$kT + E$	f
kT	

7) You can use the same sort of reasoning to find the fraction of particles with any number of times the extra energy **E** above the average particle energy. So the fraction of particles with extra energy of 3**E** = **f³**, the fraction of particles with extra energy of 4**E** = **f⁴**, etc.

8) To end up with an energy of 15**kT** to 30**kT**, a particle would have to get **very lucky**, so there will only be a tiny proportion of particles with this energy.

9) Because there are normally **huge numbers** of particles colliding billions of times each second, this small fraction still adds up to a large number of particles.

Edmond's activation energy was 3 cups of coffee and a whiskey chaser.

Practice Questions

Q1 Give an expression for the approximate energy per particle at a given temperature.

Q2 What range of values must the ratio ε/kT have for a process to occur at a reasonable speed?

Q3 What is activation energy?

Q4 Write down three processes that involve activation energies.

Exam Question

Q1 A sample of oil is at 300 K. It is poured into a pan and heated to a temperature of 360 K. The Boltzmann constant, $k = 1.38 \times 10^{-23}$ JK^{-1}.

(a) Estimate the average thermal energy in joules of an oil molecule at:
(i) 300 K [1 mark]
(ii) 360 K [1 mark]

(b) Explain in terms of activation energies why the oil is more fluid at 360 K than at 300 K. [2 marks]

Billions of collisions? You'd think they'd look where they're going...

It's like the particle version of Goldilocks — if the ratio between the activation energy and the average energy is too high, nothing happens. You need it just right... or bears will mock you for not knowing about activation energies.

The Boltzmann Factor

I know what you're thinking — if only there was some way to work out the ratio of particles in different energy states. Well today's your lucky day...

The **Boltzmann Factor** tells you the **Ratio** of Particles in two Energy States

1) You can say particles with different energies are in different **energy states**.
A particle with an energy of $kT + \varepsilon$ is in a higher energy state than a particle with an energy of kT.

2) You can find the ratio of particles in two different energy states:

> The **Boltzmann factor**, $e^{-\frac{\varepsilon}{kT}}$, gives the **ratio** of the **numbers of particles** in energy states ε joules apart.

3) For an activation energy of ε, processes start happening **quickly** when ε/kT is between 15 and 30, so try these values in the **Boltzmann factor**.

4) For ε/kT = 15 the Boltzmann factor is ~10^{-7}, and for ε/kT = 30 it's only ~10^{-13}.

5) That means that only about **one in 10^{13}** to **one in 10^7** particles have **enough energy** to overcome the activation energy.

6) That might sound like a **tiny** proportion, but you have to remember how **fast** these particles are moving.

7) Think about a reaction between two gases: gas particles collide about **10^9 times every second**. Every time there's a collision, there's an 'attempt' at the reaction, so even with **so few** particles having enough energy, the reaction can happen in a matter of **seconds**.

Bob and Rodger were both enjoying being in a low energy state.

The **Boltzmann Factor** varies with **Temperature**

For any particular **reaction**, the values of ε (activation energy) and k (Boltzmann's constant) are **fixed**. This means that the **only** thing that will change the **Boltzmann factor** is the **temperature**.

If you plot a **graph** of the **Boltzmann factor** against **temperature** you get an s-shaped curve like the ones below.

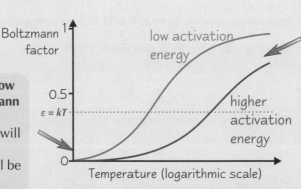

This **shape** shows that at **low temperatures**, the **Boltzmann factor** is also very **low**, so **very few** (if any) particles will have sufficient **energy** to **react** and the reaction will be really **slow**.

At **high temperatures**, the **Boltzmann factor approaches 1**, so nearly **all** the **particles** will have enough **energy** to **react** and the **reaction** will be really **fast**.

In between, the Boltzmann factor **increases rapidly** with **temperature**. So a **small increase** in **temperature** can make a **big difference** to the rate.

The Boltzmann factor and **rate** of a reaction both vary with **temperature**, and it is a **reasonable approximation** to say:

> The **rate** of a reaction with **activation energy** ε is proportional to the **Boltzmann factor**, $e^{-\frac{\varepsilon}{kT}}$.

The Boltzmann Factor

You can use the Boltzmann Factor to Describe the **Rate of a Reaction**

So, now you can use all that lovely knowledge to answer **exam questions** about the rate of a reaction, just like this one.

Example Ben has a flask filled with liquid X. The energy, ε, binding one molecule in liquid X is 0.4 eV.

(a) Estimate the average energy of the liquid molecules at 50 K. [$k = 1.38 \times 10^{-23}$ JK^{-1}]

> Approx. average energy $= kT = 1.38 \times 10^{-23} \times 50 = \mathbf{6.9 \times 10^{-22}}$ **J**

(b) Calculate the ratio ε/kT for one molecule escaping the liquid at a temperature of 50 K.

> $\varepsilon = 0.4$ eV
> $kT = 1.38 \times 10^{-23} \times 50 = 6.9 \times 10^{-22}$ J
> Convert this energy from joules into electronvolts: $6.9 \times 10^{-22} \div 1.6 \times 10^{-19} = 4.31 \times 10^{-3}$ eV
> So $\varepsilon/kT = 0.4 \div 4.31 \times 10^{-3} = \mathbf{92.8}$

(c) Calculate the Boltzmann factor for the liquid molecules at this temperature.

> $e^{-\varepsilon/kT} = e^{-92.8} = \mathbf{5.0 \times 10^{-41}}$ (1 s.f.)

(d) Ben says, 'The ratio of ε/kT is very high at 50 K, so there will be a rapid rate of evaporation of the liquid X at 50 K.' Do you agree? Use your answers to parts (b) and (c) to explain your answer.

> I disagree. A high ratio of activation energy to average particle energy means the activation energy is a lot higher than the average energy of the liquid molecules. For a good rate of reaction the ε/kT ratio should be lower (around 15-30). The Boltzmann factor at 50 K is extremely small. The rate of evaporation will be proportional to the Boltzmann factor, so the rate of evaporation of the liquid will be extremely slow.

Practice Questions

Q1 What is the Boltzmann factor?

Q2 How is it related to the rate of a reaction?

Q3 Describe the rate of reaction if the Boltzmann factor is almost 1.

Q4 Describe the relationship between the Boltzmann factor and temperature for a reaction.

Exam Question

Q1 An open fish tank is in a room at a temperature of 300 K [$k = 1.38 \times 10^{-23}$ JK^{-1}].

(a) Calculate the approximate average energy of one of the water molecules in the tank. [1 mark]

(b) Water molecules are joined together by two hydrogen bonds. The energy needed to break each bond is 3.2×10^{-20} J. Calculate the energy, ε, that a water molecule needs in order to evaporate. [1 mark]

(c) Use your answers to parts (a) and (b) to find the ratio ε/kT. [1 mark]

(d) Using the ratio ε/kT, explain why the water in the tank must be topped up regularly. [3 marks]

The Boltzmann Factor — not as much fun as the X Factor...

You can think of the Boltzmann factor as a tool in finding the probability of a particle having a certain energy, or the fraction of particles that have that energy. Or you could think of it as a big pair of grandma pants with pink polka dots — up to you.

Magnetic Fields and Motors

Magnetic fields are not just about making pretty patterns with iron filings — they're really useful too. They help distribute electricity to people's houses and allow lots of the devices in the home to work too. Magnetic fields, we salute you.

A **Magnetic Field** is a **Region** Where a **Force** is Exerted on **Magnetic Materials**

Magnetic fields can be represented by **field lines**. Field lines go from **north to south**. The **strength** of a magnetic field is represented by how **tightly packed** the lines are — the **closer** together the lines, the **stronger** the field.

The **field lines** around a **bar magnet**, or between a pair of magnets, have characteristic shapes:

At a <u>neutral point</u> magnetic fields <u>cancel out</u>.

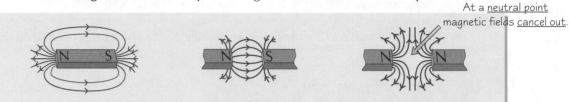

There is a **Magnetic Field** Around a **Wire** Carrying **Electric Current**

1) The **direction** of a magnetic **field** around a current-carrying wire can be worked out with the **right-hand rule**.

2) You also need to learn these diagrams for a **single coil** and a **solenoid**.

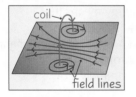

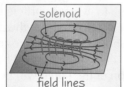

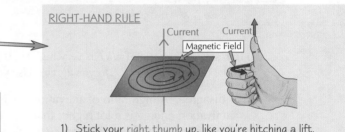

RIGHT-HAND RULE

1) Stick your <u>right thumb</u> up, like you're hitching a lift.
2) If your <u>thumb</u> points in the direction of the <u>current</u>...
3) ...your curled <u>fingers</u> point in the direction of the <u>field</u>.

A **Wire** Carrying a **Current** in a **Magnetic Field** will **Experience** a **Force**

1) If you put a **current-carrying wire** into an **external** magnetic field (e.g. between two magnets), the field around the wire and the field from the magnets **interact**. The field lines from the magnet **contract** to form a **'stretched catapult'** effect where the flux lines are closer together.

2) This causes a **force** on the wire.

3) If the current is **parallel** to the flux lines, **no force** acts.

4) The **direction** of the force is always **perpendicular** to both the **current** direction and the **magnetic field**.

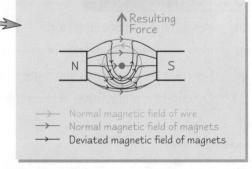

→ Normal magnetic field of wire
→ Normal magnetic field of magnets
→ Deviated magnetic field of magnets

The **Direction** of the **Force** is Given by **Fleming's Left-Hand Rule**

Fleming's Left-Hand Rule

The First finger points in the direction of the uniform magnetic Field, the seCond finger points in the direction of the conventional Current. Then your thuMb points in the direction of the force (in which Motion takes place).

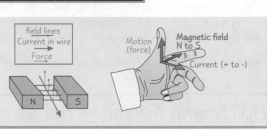

Magnetic Fields and Motors

The Size of the Force is Given by F = BIl

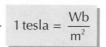

1) The size of the **force**, **F**, on a current-carrying wire at right-angles to a magnetic field is proportional to the **current**, **I**, the **length of wire** in the field, **l**, and the **strength of the magnetic field**, **B**.
This gives the equation:

$$F = BIl$$

2) In this equation, the **magnetic field strength**, **B**, is defined as:

> The **force** on **one metre** of wire carrying a **current** of **one amp** at **right angles** to the **magnetic field**.

3) **Magnetic field strength** is also called **flux density** and it's measured in **teslas, T**. ⟶ $1 \text{ tesla} = \dfrac{\text{Wb}}{\text{m}^2}$

It helps to think of flux density as the number of flux lines (measured in webers (Wb), see p 48) per unit area.

4) Magnetic field strength is a **vector** quantity with both a **direction** and **magnitude**.

The Forces on a Loop can be Used to Make a Motor

1) If a **current-carrying loop** is placed in a **magnetic field**, the **forces** on the side arms will tend to make the loop **rotate**.

2) By using a **split-ring commutator**, the current in a loop can be **reversed** each time the loop becomes **vertical** (i.e. every **half turn**).

3) This allows the loop to **rotate steadily** — which is otherwise known as a **motor**.

4) Motors are used in loads of **electronic devices** — e.g. hairdryers, vacuum cleaners, CD players.

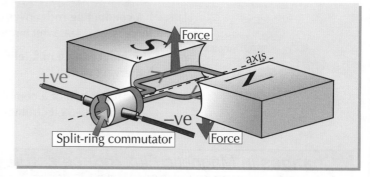

5) This isn't the only way you can use magnetism to make a motor. **Induction motors** operate by altering the magnetic field around a coil of wire that is free to move, which induces a current in the wire, causing it to rotate. Don't worry about it too much for now — the thrills of electromagnetic induction await you over the page.

Practice Questions

Q1 Describe why a current-carrying wire at right angles to an external magnetic field will experience a force.

Q2 Write down the equation you would use to find the force on a current-carrying wire.

Q3 Sketch the magnetic fields around a long straight current-carrying wire and a solenoid. Show the direction of the current and magnetic field on each diagram.

Q4 A copper bar can roll freely on two copper supports, as shown in the diagram. ⟶ When current is applied in the direction shown, which way will the bar roll?

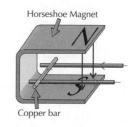

Horseshoe Magnet

Copper bar

Exam Question

Q1 A square loop of wire carrying a current of 3 A is within a magnetic field of strength 2×10^{-5} T. Each side is 4 cm long. Side A is at right angles to the magnetic field. Side B runs parallel to the magnetic field.

(a) Here is a list of forces: 2.0×10^{-6} N 4.8×10^{-6} N 2.4×10^{-6} N   0 N
State which value in this list is the force on side B of the loop.
[1 mark]

(b) Calculate the magnitude of the force on side A of the loop.
[2 marks]

I revised the right-hand rule by the A69 and ended up in Newcastle...

Fleming's left-hand rule is the key to this section — so make sure you know how to use it and understand what it all means. Remember that the direction of the magnetic field is from N to S, and that the current is from +ve to −ve — this is as important as using the correct hand. You need to get those right or it'll all go to pot...

Electromagnetic Induction

Producing electricity by waggling a wire about in a magnetic field sounds like monkey magic — but it's real physics...

Think of the **Magnetic Flux** as the Total **Number** of **Field Lines**...

1) Magnetic field strength, or **magnetic flux density**, *B*, measures the **strength** of the magnetic field **per unit area**.

2) So, the total **magnetic flux**, Φ, passing through an **area**, *A*, perpendicular to a **magnetic field**, *B*, is defined as:

$$\Phi = BA$$

(The unit of Φ is the weber, Wb. 1 tesla = 1 Wb m⁻².)

3) When you move a **coil** in a magnetic field, the size of the e.m.f. induced depends on the **magnetic flux** passing through the coil, Φ, and the **number of turns** on the coil, *N*. The product of these is called the **flux linkage**, Φ*N*.

Example

Area, *A* = 3 m²

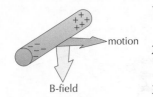

Flux density, *B* = 4 × 10⁻³ T
(flux per unit area)

$$\Phi = BA$$
$$= 4 \times 10^{-3} \times 3$$
$$= 12 \times 10^{-3} \text{ Wb}$$

Charges Accumulate on a Conductor Moving Through a Magnetic Field

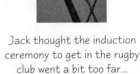

motion

B-field

1) If a **conducting rod** moves through a magnetic field its **electrons** will experience a **force** (see p. 46), which means that they will **accumulate** at one end of the rod.

2) This **induces** an **e.m.f.** (electromotive force) across the ends of the rod exactly as a **battery** would.

3) If the rod is part of a complete **circuit**, then an induced **current** will **flow** through it — this is called **electromagnetic induction**.

Changes in Magnetic Flux Induce an Electromotive Force

1) An **electromotive force** (e.m.f.) is **induced** whenever there is **relative motion** between a **conductor** and a **magnet**.

2) The **conductor** can **move** and the **magnetic field** stay **still** or the **other way round** — you get an e.m.f. either way.

3) An **e.m.f.** is **produced** whenever **lines of force** (flux) are **cut**.

4) **Flux cutting** always induces e.m.f. but will only **induce** a **current** if the **circuit** is complete.

Jack thought the induction ceremony to get in the rugby club went a bit too far...

These Results are Summed up by Faraday's Law...

FARADAY'S LAW: The **induced e.m.f.** is **directly proportional** to the **rate of change of flux linkage.**

1) **Faraday's law** can be written as: Induced e.m.f. $= \dfrac{\text{flux change}}{\text{time taken}} = \dfrac{\Delta(\Phi N)}{\Delta t}$

2) The **size** of the e.m.f. is shown by the **gradient** of a graph of Φ*N* against time.

3) The **area under** the graph of e.m.f. against time gives the **flux change**.

Φ*N* ↑ e.m.f. = gradient

time

e.m.f. ↑

Φ*N* = area

time

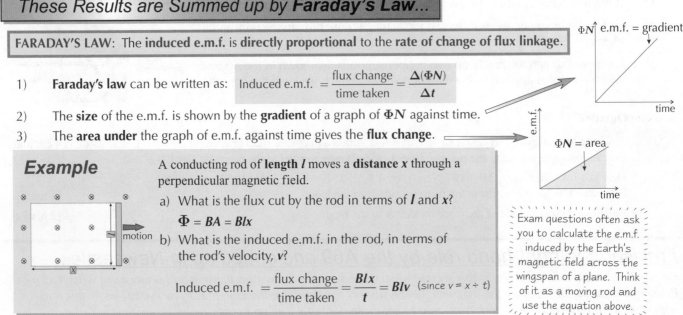

Example

A conducting rod of **length** *l* moves a **distance** *x* through a perpendicular magnetic field.

a) What is the flux cut by the rod in terms of *l* and *x*?

$$\Phi = BA = Blx$$

motion

b) What is the induced e.m.f. in the rod, in terms of the rod's velocity, *v*?

Induced e.m.f. $= \dfrac{\text{flux change}}{\text{time taken}} = \dfrac{Blx}{t} = Blv$ (since *v* = *x* ÷ *t*)

Exam questions often ask you to calculate the e.m.f. induced by the Earth's magnetic field across the wingspan of a plane. Think of it as a moving rod and use the equation above.

Electromagnetic Induction

The **Direction** of the **Induced E.m.f.** and **Current** are given by **Lenz's Law**...

> **LENZ'S LAW:** The **induced e.m.f.** is always in such a **direction** as to **oppose** the **change** that caused it.

1) **Lenz's law** and **Faraday's law** can be **combined** to give one formula that works for both:

$$\text{Induced e.m.f.} = -\frac{d(\Phi N)}{dt}$$

2) The **minus sign** shows the direction of the **induced e.m.f.**

3) The idea that an induced e.m.f. will **oppose** the change that caused it agrees with the principle of the **conservation of energy** — the **energy used** to pull a conductor through a magnetic field, against the **resistance** caused by magnetic **attraction**, is what **produces** the **induced current**.

4) **Lenz's law** can be used to find the **direction** of an **induced e.m.f.** and **current** in a conductor travelling at right angles to a magnetic field...

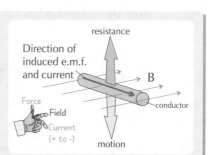

1) **Lenz's law** says that the **induced e.m.f.** will produce a force that **opposes** the motion of the conductor — in other words a **resistance**.

2) Using **Fleming's left-hand rule** (see p.46), point your thumb in the direction of the force of **resistance** — which is in the **opposite direction** to the motion of the conductor.

3) Your **second finger** will now give you the direction of the **induced e.m.f.**

4) If the conductor is **connected** as part of a **circuit**, a current will be induced in the **same direction** as the induced e.m.f.

Practice Questions

Q1 What is the difference between magnetic flux density, magnetic flux and magnetic flux linkage?

Q2 State Faraday's and Lenz's laws.

Q3 Explain how you can find the direction of an induced e.m.f. in a copper bar moving at right angles to a magnetic field.

Exam Questions

Q1 A coil of area 0.23 m² is placed at right angles to a magnetic field of 2×10^{-3} T.
   (a) Calculate the magnetic flux passing through the coil. [2 marks]
   (b) The coil has 150 turns. Calculate the magnetic flux linkage in the coil. [2 marks]
   (c) Over a period of 2.5 seconds the magnetic flux is reduced uniformly to 3.5×10^{-4} Wb.
       Calculate the size of the e.m.f. induced across the ends of the coil. [2 marks]

Q2 A 10 cm long conducting rod moves 10 cm through a perpendicular magnetic field of 0.9 T in 0.5 s.
   Which of these three statements is correct?
   A The movement will not induce a current in the rod because it is perpendicular to the magnetic field.
   B The movement will only induce a current if the rod is part of a complete circuit.
   C The rod will not cut through any magnetic flux so no current will be induced. [1 mark]

Q3 The graph shows how the flux through a coil varies over time.
   Sketch a graph to show how the induced e.m.f. in the coil
   varies over this same time period. [3 marks]

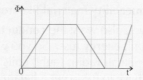

Beware — physics can induce extreme confusion...

OK... I know that might have seemed a bit scary... but the more you go through it, the more it stops being a big scary monster of doom and just becomes another couple of equations you have to remember. Plus it's one of those things that makes you sound well clever... "What did you learn today, Jim?", "Oh, just magnetic flux linkage in solenoids, Mum..."

Transformers and Alternators

Transformers are like voltage aerobics instructors. They say step up, the voltage goes up. They say step down, the voltage goes down. They say star jump, and the voltage does nothing because neither of them are alive — it's just induction.

Transformers Work by Electromagnetic Induction

1) **Transformers** are devices that make use of electromagnetic induction to **change** the size of the **voltage** for an **alternating current**. They use the principle of flux linking using two coils of wire.

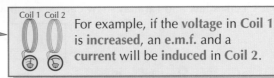

For example, if the **voltage** in **Coil 1** is **increased**, an **e.m.f.** and a **current** will be **induced** in **Coil 2**.

2) An alternating current flowing in the **primary** (or input) **coil** produces **magnetic flux**.

3) The **magnetic field** is passed through the **iron core** to the **secondary** (or output) coil, where it **induces** an alternating **voltage** of the same frequency.

4) From Faraday's law, the **induced** e.m.f.s in both the **primary** and **secondary** coils can be calculated:

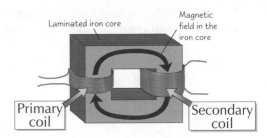

Laminated iron core
Magnetic field in the iron core
Primary coil
Secondary coil

Primary coil

$$V_p = N_p \frac{d\Phi}{dt}$$

Secondary coil

$$V_s = N_s \frac{d\Phi}{dt}$$

These can be combined to give the equation for an **ideal transformer**:

$$\frac{V_p}{V_s} = \frac{N_p}{N_s}$$

(where N is the number of turns in a coil)

5) **Step-up** transformers **increase** the **voltage** by having **more turns** on the **secondary** coil than the primary. **Step-down** transformers **reduce** the voltage by having **fewer** turns on the secondary coil.

> **Example** What is the output voltage for a transformer with a primary coil of 100 turns, a secondary coil of 300 turns and an input voltage of 230 V?
>
> $$\frac{V_p}{V_s} = \frac{N_p}{N_s} \quad \Rightarrow \quad \frac{230}{V_s} = \frac{100}{300} \quad \Rightarrow \quad V_s = \frac{230 \times 300}{100} = 690 \text{ V}$$

Permeability and Conductivity Affect Transformer Dimensions

1) Magnetic flux always forms a **closed loop**, so you can think of a magnetic field as a **magnetic circuit** (although nothing actually flows anywhere). Comparing a magnetic circuit with an electric one, you can think of the **magnetic flux** as the 'current' and the **permeance** as the 'conductance'.

2) The **permeance** of a material is the **amount of flux induced** in it for a given number of current turns that surround it. The **higher** the **permeance** of a material, the **greater** the **amount of flux** induced.

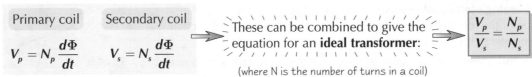

permeance, $\Lambda = \dfrac{\mu A}{L}$

3) Both permeance and conductance are **inversely proportional** to the **length** of the material, and **proportional** to the **cross-sectional area**.

Where **A** is the cross-sectional area, **L** is the length and μ is the permeability of the material.

4) When you're **designing** a transformer, you want to make the **permeance** of the core as **high** as possible to get the maximum flux induced in it. Ideally you want the core to be **short** (low L) and **fat** (high A) and made from a **high permeability** material like **iron**.

The permeability of a material, μ, is the permeance per unit cross-section of a unit length of material.

5) Unfortunately, you also want the **conductance** of the **copper coils** used on a transformer to be as **high** as possible — to limit **energy loss**. So you want to make the right **number of turns** with the **shortest** piece of wire possible, i.e. use small-radius (tight) coils. This doesn't really work when you have a fat core to wrap them around, so you have to try to get a **balance** in **dimensions** to get the **best** overall transformer performance.

6) Unlike an electric circuit, a magnetic circuit will still work if there's an **air** (or vacuum) **gap** in it. So in the case of transformers, if there's an air gap in an otherwise iron core, magnetic flux stills 'flows' around the circuit. But because air has a very **low permeability** compared to iron, the amount of flux in the circuit would be **dramatically lower** than without the air gap.

Transformers and Alternators

Transformers are **Not 100% Efficient**

1) If a transformer was **100% efficient** the **power in** would **equal** the **power out**. However, in practice there will be **small losses** of **power** from the transformer, mostly in the form of **heat**.

2) **Heat** can be produced by **eddy currents** in the transformer's iron core — currents **induced** by the changing magnetic flux in the core. This effect is reduced by **laminating** the core with layers of **insulation**.

3) Heat is also generated by **resistance** in the coils — to minimise this, **thick copper wire** is used, which has a **low resistance**.

Transformers are an **Important** Part of the **National Grid**...

1) **Electricity** from power stations is sent round the country in the **national grid** at the **lowest** possible current, because **losses** due to the **resistance** in the cables are proportional to I^2 — so if you double the transmitted current, you quadruple the power lost.

2) Since **power** = **current** × **voltage**, a **low current** means a **high voltage**.

3) **Transformers** allow us to **step up** the voltage to around **400 000 V** for **transmission** through the national grid, and then **reduce** it again to **230 V** for domestic use.

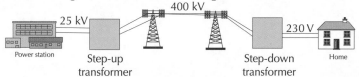

... robots in disguise

An **Alternator** is a **Generator** of **Alternating Current**

1) **Generators**, or dynamos, **convert** kinetic energy into **electrical energy** — they **induce** an electric **current** by **rotating** a **coil** in a magnetic field.

2) A simple **alternator** looks similar to a **motor** but with **slip rings** and **brushes** instead of a split-ring commutator.

3) The output **voltage** and **current** change direction with every **half rotation** of the coil, producing **alternating current** (**AC**).

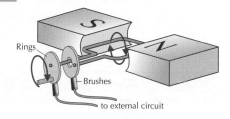

Practice Questions

Q1 Draw a diagram of a simple transformer. What is meant by a step-down transformer?

Q2 Describe how you could minimise heat loss in a transformer core due to eddy currents.

Q3 What is meant by the permeance of a material? Describe what considerations you need to take into account when designing the shape of a transformer.

Exam Questions

Q1 An ideal transformer with 150 turns in the primary coil has an input voltage of 9 V.
  (a) How many turns are needed in the secondary coil to step up the voltage to 45 V?   [2 marks]
  (b) The input current for the transformer is 1.5 A. Calculate the output current of the transformer.   [2 marks]
  (c) Calculate the efficiency of the transformer when its power output is 10.8 W.   [2 marks]

Q2 Using Lenz's law, explain why eddy currents are set up in real transformers, and why they alter the flux in the transformer core.
(Marks will be given for the quality of your written communication.)   [4 marks]

Arrrrrrrrrggggggggghhhhhhhh...

Breathe a sigh of relief, pat yourself on the back and make a brew — well done, you've reached the end of the section. That was pretty nasty stuff, but don't let all of those equations get you down — once you've learnt the main ones and can use them blindfolded, even the trickiest looking exam question will be a walk in the park...

Electric Fields

*Electric fields can be attractive or repulsive, so they're different from gravitational ones. It's all to do with **charge**.*

There is an **Electric Field** around a **Charged Object**

Any object with **charge** has an **electric field** around it — the region where it can attract or repel other charges.

1) Electric charge, **Q**, is measured in **coulombs** (C) and can be either positive or negative.

2) **Oppositely** charged particles **attract** each other. **Like** charges **repel**.

3) If a **charged object** is placed in an electric field, then it will experience a **force**.

You can **Calculate Forces** using **Coulomb's Law**

You'll need **Coulomb's law** to work out **F** — the force of attraction or repulsion between two point charges...

COULOMB'S LAW:

$$F = \frac{kQ_1Q_2}{r^2} \quad \text{where} \quad k = \frac{1}{4\pi\varepsilon}$$

ε ("epsilon") = permittivity of material between charges
Q_1 and Q_2 are the charges
r is the distance between Q_1 and Q_2

If the charges are **opposite** then the force is **attractive**. **F** will be **negative**.

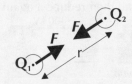

If Q_1 and Q_2 are **like** charges then the force is **repulsive**, and **F** will be **positive**.

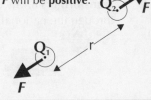

1) The force on Q_1 is always **equal** and **opposite** to the force on Q_2.

2) It's an **inverse square law**. Again. The further apart the charges are, the weaker the force between them.

3) The size of the force **F** also depends on the **permittivity**, ε, of the material between the two charges. For free space, the permittivity is $\varepsilon_0 = 8.85 \times 10^{-12} \, \text{C}^2\text{N}^{-1}\text{m}^{-2}$.

Electric Field Strength is Force per Unit Charge

Electric field strength, **E**, is defined as the **force per unit positive charge** — the force that a charge of +1 C would experience if it was placed in the electric field.

$$E = \frac{F}{q}$$

F is the force on a 'test' charge **q**.

1) **E** is a **vector** pointing in the **direction** that a **positive charge** would **move**.

2) The units of **E** are **newtons per coulomb** (NC^{-1}).

3) Field strength depends on **where you are** in the field.

4) A **point charge** — or any body that behaves as if all its charge is concentrated at the centre — has a **radial** field.

In a **Radial Field**, E is **Inversely Proportional** to r^2

1) **E** is the force per unit charge that a small, positive 'test' charge, **q**, would feel at different points in the field. In a **radial field**, **E** depends on the distance **r** from the point charge **Q**...

$$E = \frac{kQ}{r^2} \quad \left(k = \frac{1}{4\pi\varepsilon}\right)$$

For a **positive Q**, the small positive 'test' charge **q** would be **repelled**, so the field lines point **away** from Q.

For a **negative Q**, the small positive charge **q** would be **attracted**, so the field lines point **towards** Q.

2) It's another **inverse square law** — $E \propto \dfrac{1}{r^2}$

3) Field strength **decreases** as you go **further away** from **Q** — on a diagram, the **field lines** get **further apart**.

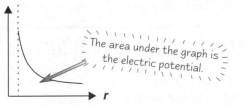

The area under the graph is the electric potential.

Electric Fields

A **Charge** in an Electric Field has **Electric Potential Energy**

The electric potential energy, $E_{electric}$, is the **work** that would need to be done to move a small charge, q, from infinity to a distance r away from a point charge, Q...

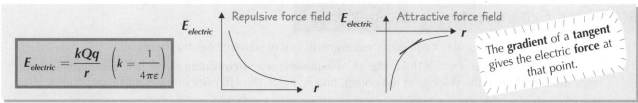

$$E_{electric} = \frac{kQq}{r} \quad \left(k = \frac{1}{4\pi\varepsilon}\right)$$

Repulsive force field / Attractive force field

The **gradient** of a tangent gives the electric **force** at that point.

1) At an **infinite** distance from Q, a charged particle q would have **zero potential energy.**

2) In a **repulsive** force field (e.g. Q and q are both positive) you have to **do work** against the repulsion to bring q closer to Q. The charge q **gains** potential energy as r **decreases**.

3) In an **attractive** field (e.g. Q negative and q positive) the charge q **gains** potential energy as r **increases**.

Electric Potential is Potential Energy per Unit Charge

Electric potential, V, is electric **potential energy** per **unit positive charge**...

$V = \dfrac{E_{electric}}{q}$ and substituting for $E_{electric}$ gives

$$V = \frac{kQ}{r} \quad \left(k = \frac{1}{4\pi\varepsilon}\right)$$

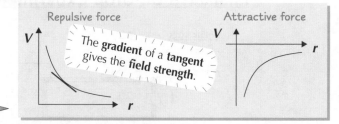

Repulsive force / Attractive force

The **gradient** of a tangent gives the **field strength**.

1) V is measured in **volts**.

2) As with E, V is **positive** when the force is **repulsive**, and **negative** when the force is **attractive**...

Field Strength is the Same Everywhere in a Uniform Field

A **uniform field** can be produced by connecting two **parallel plates** to the opposite poles of a battery.

1) Field strength E is the **same** at **all points** between the two plates and is...

$$E = \frac{V}{d}$$

V is the **potential difference** between the plates
d is the distance between them

2) E can be measured in volts per metre (Vm^{-1})

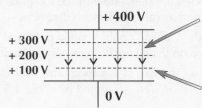

+ 400 V
+ 300 V
+ 200 V
+ 100 V
0 V

The **lines of force** are **parallel** to each other.

The **equipotential surfaces** are **parallel** to the **plates**, and **perpendicular** to the **field lines**.

Practice Questions

Q1 Draw the electric field lines due to a positive charge, and due to a negative charge.
Q2 Write down Coulomb's law.

Exam Questions

Q1 The diagram shows two electric charges with equal but opposite charge, Q.
Draw electric field lines to show the electric field in the area surrounding the charges. $+Q$ $-Q$ [3 marks]

Q2 Find the electric field strength at a distance of 1.75×10^{-10} m from a 1.6×10^{-19} C point charge. [2 marks]

Q3 (a) Two parallel plates are separated by an air gap of 4.5 mm. The plates are connected to a 1500 V dc supply. What is the electric field strength between the plates? Give a suitable unit and state the direction of the field. [3 marks]

(b) The plates are now pulled further apart so that the distance between them is doubled.
The electric field strength remains the same. What is the new voltage between the plates? [2 marks]

Electric fields — one way to roast beef...

At least you get a choice here — uniform or radial, positive or negative, attractive or repulsive, chocolate or strawberry...

Millikan's Oil-Drop Experiment

You'll probably know that the fundamental unit of charge, i.e. the charge on an electron, is 1.6 × 10⁻¹⁹ C, and that all other charges are always exact multiples of this value. But, at the beginning of the 20th century, many physicists thought that charge was a continuous variable, so could take any value at all — Millikan showed this wasn't true.

Millikan's Experiment used Stoke's Law

1) Before you start thinking about Millikan's experiment, you need a bit of **extra theory**.

2) When you drop an object into a fluid, like air, it experiences a **viscous drag** force. This force acts in the **opposite direction** to the velocity of the object, and is due to the **viscosity** of the fluid.

3) You can calculate this viscous force on a spherical object using **Stoke's law**:

$$F = 6\pi\eta rv$$

where η is the viscosity of the fluid, r is the radius of the object and v is the velocity of the object.

Millikan's Experiment — the Basic Set-Up

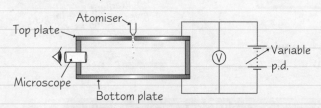

Millikan's Oil-Drop Experiment — Apparatus

Millikan's Set-Up

Atomiser

Top plate

Microscope

Bottom plate

V

Variable p.d.

1) The **atomiser** created a **fine mist** of oil drops that were **charged** by **friction** as they left the atomiser (positively if they lost electrons, negatively if they gained electrons).

2) Some of the drops fell through a **hole** in the top plate and could be viewed through the **microscope**. (The eyepiece carried a **scale** to measure distances — and so **velocities** — accurately.)

3) When he was ready, Millikan could apply a **potential difference** between the two plates, producing a **field** that exerted a **force** on the charged drops. By **adjusting** the p.d., he could vary the strength of the field.

To give you a feel for the **size** of the apparatus, Millikan's plates were circular, with a diameter of about the width of this page. They were separated by about 1.5 cm.

Before the Field is Switched on, there's only Gravity and the Viscous Force

1) With the electric field turned off, the forces acting on each oil drop are:

a) the **weight** of the drop — acting downwards

b) the **viscous force** from the air — acting upwards

Millikan had to take account of things like upthrust as well, but you don't have to worry about that — keep it simple.

2) The drop will reach **terminal velocity** (i.e. it will stop accelerating) when these two forces are equal. So, from Stoke's law (see above):

$$mg = 6\pi\eta rv$$

3) Since the **mass** of the drop is the **volume** of the drop multiplied by the **density**, ρ, of the oil, this can be rewritten as:

$$\frac{4}{3}\pi r^3 \rho g = 6\pi\eta rv \quad \Rightarrow \quad r^2 = \frac{9\eta v}{2\rho g}$$

Millikan measured η and ρ in separate experiments, so he could now calculate r — ready to be used when he switched on the electric field...

Millikan's Oil-Drop Experiment

Then he Turned On the Electric Field...

1) The field introduced a **third major factor** — an **electric force** on the drop.

2) Millikan adjusted the applied p.d. until the drop was **stationary**. Since the **viscous force** is proportional to the **velocity** of the object, once the drop stopped moving, the viscous force **disappeared**.

3) Now the only two forces acting on the oil drop were:

    a) the **weight** of the drop — acting downwards

    b) the force due to the **uniform electric field** — acting upwards

4) The **electric force** and the **p.d.** are related to the **electric field strength** by the equations from pages 52 and 53:

$$E = \frac{F}{q} \qquad\qquad E = \frac{V}{d}$$

5) Combining these equations gives:

$$F = \frac{QV}{d}$$

where Q is the charge on the oil drop, V is the p.d. between the plates and d is the distance between the plates.

6) Since the drop is **stationary**, this electric force must be equal to the weight, so:

$$\frac{QV}{d} = \frac{4}{3}\pi r^3 \rho g$$

The first part of the experiment gave a value for r, so the **only unknown** in this equation is Q.

7) So Millikan could find the **charge on the drop**, and repeated the experiment for hundreds of drops. The charge on any drop was always a **whole number multiple** of -1.6×10^{-19} C.

These Results Suggested that Charge was Quantised

1) This result was **really significant**. Millikan concluded that charge can **never exist** in **smaller** quantities than 1.6×10^{-19} C. He assumed that this was the **charge** carried by an **electron**.

2) Later experiments confirmed that **both** these things are true.

> Charge is "**quantised**". It exists in discrete "packets" of size **1.6×10^{-19} C** — the **fundamental unit of charge.** This is the size of the charge carried by **one electron**.

Practice Questions

Q1 List the forces that act on the oil drop in Millikan's experiment:
(a) with the drop drifting downwards at terminal velocity but with no applied electrical field,
(b) when the drop is stationary, with an electrical field applied.

Exam Question

Q1 An oil drop of mass 1.63×10^{-14} kg is held stationary in the space between two charged plates 3.00 cm apart. The potential difference between the plates is 5000 V. The density of the oil used is 880 kgm^{-3}.

(a) Describe the relative magnitude and direction of the forces acting on the oil drop. [2 marks]

(b) Calculate the charge on the oil drop using $g = 9.81$ Nkg^{-1}.
Give your answer in terms of e, the charge on an electron. [3 marks]

The electric field is switched off and the oil drop falls towards the bottom plate.

(c) Explain why the oil drop reaches terminal velocity as it falls. [3 marks]

(d) Calculate the terminal velocity of the oil drop using $\eta = 1.84 \times 10^{-5}$ kgm^{-1}s^{-1}. [3 marks]

So next time you've got a yen for 1.59×10^{-19} coulombs — tough...

This was a huge leap. Along with the photoelectric effect this experiment marked the beginning of quantum physics. The world wasn't ruled by smooth curves any more — charge now jumped from one allowed step to the next... gosh.

Charged Particles in Magnetic Fields

Charged particles can be deflected by magnetic fields because the field exerts a force on the particles. This is the same effect that you saw in Unit 5: Section 1 where a magnetic field exerted a force on a current-carrying wire.

Forces Act on Charged Particles in Magnetic Fields

Electric current in a wire is caused by the **flow** of negatively **charged** electrons. These charged particles are affected by **magnetic fields** — so a current-carrying wire experiences a **force** in a magnetic field (see pages 46–47).

1) The equation for the **force** exerted on a **current-carrying wire** in a **magnetic field** perpendicular to the current is:

   Equation 1: $F = BIl$

2) To see how this relates to **charged particles** moving through a wire, you need to know that electric **current**, I, is the flow of **charge**, q, per unit **time**, t.

   $I = \dfrac{q}{t}$

   In many exam questions, q is the size of the charge on the electron, which is 1.6×10^{-19} coulombs.

3) A charged particle which moves a **distance** l in **time** t has a **velocity**, v:

   $$v = \frac{l}{t} \Rightarrow t = \frac{l}{v}$$

4) So, putting the two equations **together** gives the **current** in terms of the **charge** flowing through the **wire**:

   Equation 2: $I = \dfrac{qv}{l}$

5) Putting **equation 2** back into **equation 1** gives the **electromagnetic force** on the wire as:

   $F = qvB$

6) You can use this equation to find the **force** acting on a **single charged particle moving through a magnetic field**.

   ### Example
   What is the force acting on an electron travelling at 2×10^4 ms^{-1} through a uniform magnetic field of strength 2 T?
   (The magnitude of the charge on an electron is 1.6×10^{-19} C.)

   $F = qvB$
   so, $F = 2 \times 1.6 \times 10^{-19} \times 2 \times 10^4$
   so, $F = 6.4 \times 10^{-15}$ N

Charged Particles in a Magnetic Field are Deflected in a Circular Path

1) By **Fleming's left-hand rule** the force on a **moving charge** in a magnetic field is always **perpendicular** to its **direction of travel**.

2) Mathematically, that is the condition for **circular** motion.

3) This effect is used in **particle accelerators** such as **cyclotrons** and **synchrotrons** (see pages 66–67), which use **electric and magnetic fields** to accelerate particles to very **high energies** along circular paths.

4) The **radius of curvature** of the **path** of a charged particle moving through a magnetic field gives you information about the particle's **charge** and **mass** — this means you can **identify different particles** by studying how they're **deflected**.

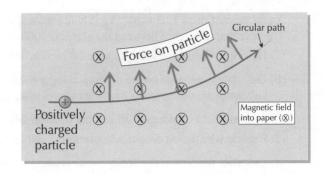

Charged Particles in Magnetic Fields

Centripetal Force Tells Us About a Particle's Path

The centripetal force and the electromagnetic force are equivalent for a charged particle travelling along a circular path.

Centripetal force   Electromagnetic force

$$F = \frac{mv^2}{r} \qquad F = qvB$$

1) For uniform circular motion **Newton's second law** gives:

$$F = \frac{mv^2}{r}$$

2) So, for a **charged particle** following a **circular** path in a **magnetic field** (where **F = qvB**):

$$qvB = \frac{mv^2}{r}$$

3) Rearranging gives:

$$r = \frac{mv}{Bq}$$

Where: m is the mass of the particle, v is its speed and r is the radius of the circular path.

Eric and Phil were *not* lost. They'd just accidentally demonstrated the path of a charge in a magnetic field.

Example   A magnetic field of strength 0.08 T is used to move an electron in a circular path with a radius of 1.8×10^{-4} m at a constant speed. Calculate the speed of the electron.

1) The force on a charged particle moving in a magnetic field, **F = qvB**

2) The particles move in a circle, so **F** gives the centripetal force $\Rightarrow qvB = \dfrac{mv^2}{r}$

3) Rearranging for v gives: $v = \dfrac{Bqr}{m}$

The charge on an electron is 1.6×10^{-19} C.

4) Substitute the values: $v = \dfrac{0.08 \times 1.6 \times 10^{-19} \times 1.8 \times 10^{-4}}{9.11 \times 10^{-31}} = \mathbf{2.5 \times 10^6}$ ms^{-1}

The mass of an electron is 9.11×10^{-31} kg.

Practice Questions

Q1   Write down an equation to calculate the force on a charged particle moving in a magnetic field.

Q2   Why are charged particles deflected in circular paths by magnetic fields?

Q3   Give two examples of how magnetic fields can be used.

Exam Questions

Q1 (a) An electron travels at a velocity of 5×10^6 ms^{-1} through a perpendicular magnetic field of 0.77 T. Find the force acting on the electron. [The charge on an electron is -1.6×10^{-19} C.]   [2 marks]

(b) Explain why the electron follows a circular path while in the field.   [1 mark]

Q2   An electron is accelerated to a velocity of 2.3×10^7 ms^{-1} by a particle accelerator. The electron moves in a circular path perpendicular to a magnetic field of 0.6 mT.

(a) Use the equations for electromagnetic and centripetal force to show that $Bq = mv \div r$.   [2 marks]

(b) Use the relation from part (a) to calculate the radius of the electron's path. [The mass of an electron is 9.11×10^{-31} kg and its charge is -1.6×10^{-19} C.]   [1 mark]

Hold on to your hats folks — this is starting to get tricky...

Basically, the main thing you need to know here is that a magnetic field will exert a force on a charged particle, making it follow a circular path. There's even a handy equation to work out the force on a charged particle moving through a magnetic field — it might not impress your friends, but it will impress the examiner, so learn it.

Scattering to Determine Structure

By firing radiation at different materials, you can take a sneaky beaky at their internal structures...

Rutherford's Experiment Disproved the Thomson Model

1) Until the early 20th century, physicists believed that the atom was a **positively charged globule** with **negatively charged electrons sprinkled** in it.

2) In Rutherford's laboratory, **Hans Geiger** and **Ernest Marsden** studied the scattering of **alpha particles** by **thin gold foils**.

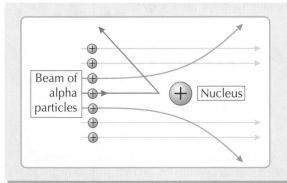

1) A **stream of alpha particles** from a radioactive source was fired at **very thin gold foil**.

2) Geiger and Marsden recorded the **number** of alpha particles scattered at **different angles**.

3) Geiger and Marsden observed that alpha particles occasionally **scatter at angles greater than 90°**. This can only be possible if they're **striking something more massive** than themselves.

Rutherford's Model of the Atom — The Nuclear Model

This experiment led Rutherford to some **important conclusions**:

1) Most of the **fast, charged alpha particles** went **straight through** the gold foil. Therefore the atom is **mostly empty space**.

2) **Some** of the alpha particles are **deflected back** through **significant angles**, so the **centre** of the atom must be **tiny** but contain **a lot of mass**. Rutherford named this the **nucleus**.

3) The **alpha particles** were **repelled**, so the **nucleus** must have **positive charge**.

4) **Atoms** are **neutral overall** so the **electrons** must be on the outside of the atom — separating one atom from the next.

Atoms are made up of Protons, Neutrons and Electrons

Inside **every atom**, there's a **nucleus** containing **protons** and **neutrons**.
Protons and **neutrons** are both known as **nucleons**. **Orbiting** this core are the **electrons**.

This is the **nuclear model** of the atom.

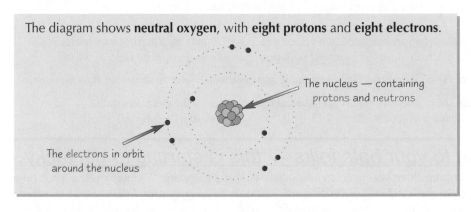

The diagram shows **neutral oxygen**, with **eight protons** and **eight electrons**.

The nucleus — containing protons and neutrons

The electrons in orbit around the nucleus

Scattering to Determine Structure

You can **Estimate** the **Closest Approach** of a **Scattered Particle**

1) When you fire an alpha particle at a gold nucleus, you know its **initial kinetic energy**.

2) An alpha particle that 'bounces back' and is deflected through 180° will have stopped a short distance from the nucleus. It does this at the point where its **electrical potential energy** (see p 53) **equals** its **initial kinetic energy**.

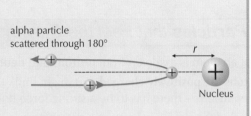
alpha particle scattered through 180°

r

Nucleus

3) It's just conservation of energy — and you can use it to find how close the particle can get to the nucleus.

$$\text{Initial K.E.} = E_{elec} = \frac{Q_{gold}\,q_{alpha}}{4\pi\varepsilon_0 r}$$

ε_o is the permittivity of free space, 8.9×10^{-12} Fm^{-1}

4) To find the charge of a nucleus you need to know the atom's **proton number**, **Z** — that tells you how many protons are in the nucleus (surprisingly).

5) A proton has a charge of **+e** (where e is the size of the charge on an electron), so the charge of a nucleus must be **+Ze**.

Example   An alpha particle with an initial kinetic energy of 6 MeV is fired at a gold nucleus. Find the closest approach of the alpha particle to the nucleus.

Initial particle energy $= 6$ MeV $= 6 \times 10^6$ eV

Convert energy into joules: $6 \times 10^6 \times 1.6 \times 10^{-19} = 9.6 \times 10^{-13}$ J

So, electrical potential energy $= E_{elec} = \dfrac{Q_{gold}\,q_{alpha}}{4\pi\varepsilon_0 r} = 9.6 \times 10^{-13}$ J at closest approach.

Rearrange to get $r = \dfrac{(+79e)(+2e)}{4\pi\varepsilon_0(9.6 \times 10^{-13})} = \dfrac{2 \times 79 \times (1.6 \times 10^{-19})^2}{4\pi \times 8.9 \times 10^{-12} \times 9.6 \times 10^{-13}} = \mathbf{3.8 \times 10^{-14}}$ **m**

Practice Questions

Q1   Describe the structure of the atom according to the nuclear model.

Q2   Explain how alpha-particle scattering shows that a nucleus is both small and positively charged.

Exam Questions

Q1   A beam of alpha particles is directed onto a very thin gold film.
(a) Explain why the majority of alpha particles are not scattered. [2 marks]
(b) Explain how alpha particles are scattered by atomic nuclei. [3 marks]

Q2   A proton is fired at a gold nucleus. It has an initial kinetic energy of 4 MeV.
Show that the closest approach of the proton to the nucleus is 2.8×10^{-14} m (to 2 s.f.).
(*The permittivity of free space, $\varepsilon_o = 8.9 \times 10^{-12}$ Fm^{-1}.*) [4 marks]

Alpha scattering — It's positively repulsive...

The important things to learn from these two pages are the nuclear model for the structure of the atom (i.e. a large mass nucleus surrounded by orbiting electrons) and how Geiger and Marsden's alpha-particle scattering experiment gives evidence that supports this model. Once you know that, take a deep breath — it's about to get a little more confusing.

Classification of Particles

There are loads of different types of particle apart from the ones you get in normal matter (protons, neutrons, etc.). They only appear in cosmic rays and in particle accelerators, and they often decay very quickly, so they're difficult to get a handle on. Nonetheless, you need to learn about a load of them and what their properties are.

Don't expect to really understand this (I don't) — you only need to learn it. Stick with it — you'll get there.

Hadrons *are* Particles *that Feel the* Strong Interaction *(e.g. Protons and Neutrons)*

1) The **nucleus** of an atom is made up from **protons** and **neutrons** (déjà vu).

2) Since the **protons** are **positively charged** you might think that the nucleus would **fly apart** with all that repulsion — there has to be a strong **force** holding the **p**'s and **n**'s together.

3) That force is called the strong interaction (who said physicists lack imagination...)

4) Not all particles can feel the strong interaction — the ones that can are called hadrons.

5) Hadrons aren't **fundamental** particles. They're made up of **smaller particles** called **quarks** (see page 64).

6) There are **two** types of **hadron** — **baryons** and **mesons** (you don't need to know about mesons though, hurrah.)

Protons *and* Neutrons *are* Baryons

1) It's helpful to think of **protons** and **neutrons** as **two versions** of the **same particle** — the **nucleon**. They just have **different electric charges**.

2) As well as **protons** and **neutrons**, there are **other baryons** that you don't get in normal matter — like **sigmas** (Σ) — they're **short-lived** and you **don't** need to **know about them** for A2 (woohoo!).

The Proton *is the* Only Stable Baryon

All baryons except protons decay to a **proton**.
Most physicists think that protons don't **decay**.

Some theories predict that protons should decay with
a very long half-life of about 10^{32} years — but there's
no experimental evidence for it at the moment.

The Number of Baryons *in a reaction is called the* Baryon Number

Baryon number is the number of baryons.
(A bit like **nucleon number** but including unusual baryons like Σ too.)
The **proton** and the **neutron** each have a baryon number **B = +1**.
The **total baryon number** in **any** particle reaction **never changes**.

Baryon and Meson felt
the strong interaction.

Leptons Don't *feel the* Strong Interaction *(e.g. Electrons and Neutrinos)*

1) **Leptons** are **fundamental particles** and they **don't** feel the **strong interaction**. The only way they can **interact** with other particles is via the **weak interaction** and gravity (and the electromagnetic force as well if they're charged).

2) **Electrons** (**e⁻**) are **stable** and very **familiar** but — you guessed it — there are also **two more leptons** called the **muon** (μ^-) and the **tau** (τ^-) that are just like **heavy electrons**.

3) **Muons** and **taus** are **unstable**, and **decay** eventually into **ordinary electrons**.

4) The **electron**, **muon** and **tau** each come with their **own neutrino**: ν_e, ν_μ and ν_τ. ν is the Greek letter "nu".

5) **Neutrinos** have **zero** or **almost zero mass** and **zero electric charge** — so they don't do much. In fact, a neutrino can **pass right through the Earth** without **anything** happening to it.

Classification of Particles

You Have to **Count** the **Three Types** of Lepton **Separately**

Each lepton is given a **lepton number** of **+1**, but the **electron**, **muon** and **tau** types of lepton have to be **counted separately**.

You get **three different** lepton numbers: L_e, L_μ and L_τ.

Name	Symbol	Charge	L_e	L_μ	L_τ
electron	e^-	–1	+1	0	0
electron neutrino	ν_e	0	+1	0	0
muon	μ^-	–1	0	+1	0
muon neutrino	ν_μ	0	0	+1	0
tau	τ^-	–1	0	0	+1
tau neutrino	ν_τ	0	0	0	+1

Like the baryon number, the lepton number is just the number of leptons.

Neutrons Decay into Protons

The **neutron** is an **unstable particle** that **decays** into a **proton**. (But it's much more stable when it's part of a nucleus.) It's really just an **example** of β^- decay:

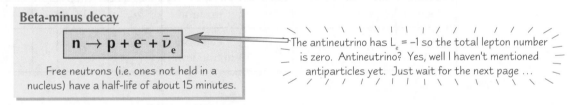

Beta-minus decay

$$n \rightarrow p + e^- + \bar{\nu}_e$$

Free neutrons (i.e. ones not held in a nucleus) have a half-life of about 15 minutes.

The antineutrino has $L_e = -1$ so the total lepton number is zero. Antineutrino? Yes, well I haven't mentioned antiparticles yet. Just wait for the next page ...

Practice Questions

Q1 List the differences between a hadron and a lepton.

Q2 Which is the only stable baryon?

Q3 A particle collision at CERN produces 2 protons, 3 muons and 1 neutron. What is the total baryon number of these particles?

Q4 Which two particles have lepton number $L_\tau = +1$?

Exam Question

Q1 This equation shows the beta decay of a neutron.

$$n \rightarrow p + e^- + X$$

(a) Name the missing decay product of the neutron, **X**. [1 mark]

(b) Explain why this decay cannot be due to the strong interaction. [2 marks]

Go back to the top of page 60 — do not pass GO, do not collect £200...

Do it. Go back and read it again. I promise — read these pages about 3 or 4 times and you'll start to see a pattern. There are hadrons that feel the force, leptons that don't. Hadrons are either baryons or mesons, and they're all weird except for those well-known baryons: protons and neutrons. There are loads of leptons, including good old electrons.

Antiparticles

More stuff that seems to laugh in the face of common sense — but actually, antiparticles help to explain a lot in particle physics... (Oh, and if you haven't read pages 60 and 61 yet then go back and read them now — no excuses, off you go...)

Antiparticles were Predicted Before they were Discovered

When **Paul Dirac** wrote down an equation obeyed by **electrons**, he found a kind of **mirror image** solution.

1) It predicted the existence of a particle like the **electron** but with **opposite electric charge** — the **positron**.

2) The **positron** turned up later in a cosmic ray experiment. Positrons are **antileptons** so $L_e = -1$ for them. They have **identical mass** to electrons but they carry a **positive** charge.

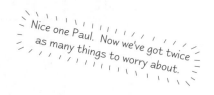
Nice one Paul. Now we've got twice as many things to worry about.

Every Particle has an Antiparticle

Each particle type has a **corresponding antiparticle** with the **same mass** but with **opposite charge**. For instance, an **antiproton** is a **negatively charged** particle with the same mass as the **proton**.

Even the shadowy **neutrino** has an antiparticle version called the **antineutrino** — it doesn't do much either.

Particle	Symbol	Charge	B	L_e	Antiparticle	Symbol	Charge	B	L_e
proton	p	+1	+1	0	antiproton	$\bar{p}$	−1	−1	0
neutron	n	0	+1	0	antineutron	$\bar{n}$	0	−1	0
electron	e	−1	0	+1	positron	e^+	+1	0	−1
electron neutrino	ν_e	0	0	+1	electron antineutrino	$\bar{\nu}_e$	0	0	−1

You can Create Matter and Antimatter from Energy

You've probably heard about the **equivalence** of energy and mass. It all comes out of Einstein's special theory of relativity. **Energy** can turn into **mass** and **mass** can turn into **energy** if you know how — all you need is one fantastic and rather famous formula.

$$E = mc^2$$

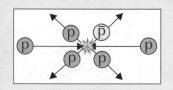

It's a good thing this doesn't randomly happen all the time or else you could end up with cute bunny rabbits popping up and exploding unexpectedly all over the place. Oh, the horror...

As you've probably guessed, there's a bit **more to it** than that:

> When **energy** is converted into **mass** you have to make **equal amounts** of **matter** and **antimatter**.

Fire **two protons** at each other at high speed and you'll end up with a lot of **energy** at the point of impact. This energy can form **more particles**.

If an extra **proton** is created, there has to be an **antiproton** made to go with it. It's called **pair production**.

Antiparticles

Each **Particle-Antiparticle Pair** is Produced from a **Single Photon**

Pair production only happens if **one gamma ray photon** has enough energy to produce that much mass. It also tends to happen near a **nucleus**, which helps conserve momentum.

You usually get **electron-positron** pairs produced (rather than any other pair) — because they have a relatively **low mass**.

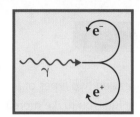

The particle tracks are curved because there's usually a magnetic field present in particle physics experiments. They curve in opposite directions because of the opposite charges on the electron and positron.

> **Example** An electron and a positron are produced from a single photon.
> Find the minimum energy of the photon. (The rest mass of an electron $m_e = 9.11 \times 10^{-31}$ kg.)
>
> The minimum energy the photon must have is enough energy to produce the particles' mass alone (the particles will have no kinetic energy).
> Energy before = Energy after.
> So the energy of the photon $\geq 2m_e c^2 = 2 \times 9.11 \times 10^{-31} \times (3 \times 10^8)^2$
> $$= 1.64 \times 10^{-13} \text{ J}$$
> $$= \textbf{1.0 MeV}$$

The **Opposite** of **Pair Production** is **Annihilation**

1) When a **particle** meets its **antiparticle** the result is **annihilation**.

2) All the **mass** of the particle and antiparticle gets converted to **energy**.

3) In ordinary matter antiparticles can only exist for a fraction of a second before this happens, so you won't see many of them.

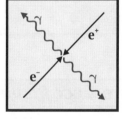

The electron and positron annihilate and their mass is converted into the energy of a pair of gamma ray photons.

Practice Questions

Q1 Which antiparticle has zero charge and a baryon number of –1?

Q2 Describe the properties of an electron antineutrino.

Q3 What is pair production? What happens when a proton collides with an antiproton?

Exam Questions

Q1 Complete the equation to show the reaction between an electron and a positron when they collide.
$e^+ + e^- \rightarrow$ State the name of this type of reaction.   [2 marks]

Q2 According to Einstein, mass and energy are equivalent.
Explain why the mass of a block of iron cannot be converted directly into energy.   [2 marks]

Q3 Give a reason why the reaction $\mathbf{p} + \mathbf{p} \rightarrow \mathbf{p} + \mathbf{p} + \mathbf{n}$ is not possible.   [1 mark]

Now stop meson around and do some work...

The idea of every particle having an antiparticle might seem a bit strange, but just make sure you know the main points — a) if energy is converted into a particle, you also get an antiparticle, b) an antiparticle won't last long before it bumps into the right particle and annihilates it with a big ba-da-boom, c) this releases the energy it took to make them to start with...

Quarks

*If you haven't read pages 60 to 63, do it now! For the rest of you — here are the **juicy bits** you've been waiting for.*
*Particle physics makes **a lot more sense** when you look at quarks. More sense than it did before anyway.*

Quarks are Fundamental Particles

Quarks are the **building blocks** for **hadrons**.

*If that first sentence doesn't make much sense to you,
<u>read pages 60-63</u> — you have been warned... twice.*

1) To make **protons** and **neutrons** you only need two types of quark — the **up** quark (**u**) and the **down** quark (**d**).

2) An extra one called the **strange** quark (**s**) lets you make more particles with a property called **strangeness**.

3) There are another three types of quark called **top**, **bottom** and **charm** (tut... physicists) that were predicted from the symmetry of the quark model. But luckily you don't have to know much about them...

The antiparticles of hadrons are made from **antiquarks**.

Quarks and Antiquarks have Opposite Properties

The **antiquarks** have **opposite properties** to the quarks — as you'd expect.

QUARKS

name	symbol	charge	baryon number	strangeness
up	u	$+\frac{2}{3}$	$+\frac{1}{3}$	0
down	d	$-\frac{1}{3}$	$+\frac{1}{3}$	0
strange	s	$-\frac{1}{3}$	$+\frac{1}{3}$	-1

ANTIQUARKS

name	symbol	charge	baryon number	strangeness
anti-up	$\bar{u}$	$-\frac{2}{3}$	$-\frac{1}{3}$	0
anti-down	$\bar{d}$	$+\frac{1}{3}$	$-\frac{1}{3}$	0
anti-strange	$\bar{s}$	$+\frac{1}{3}$	$-\frac{1}{3}$	$+1$

Baryons are Made from Three Quarks

Evidence for quarks came from **hitting protons** with **high-energy electrons**.
The way the **electrons scattered** showed that there were **three concentrations of charge** (quarks) **inside** the proton.

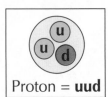

Total charge
= 2/3 + 2/3 − 1/3 = 1
Baryon number
= 1/3 + 1/3 + 1/3 = 1

Proton = **uud**

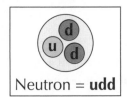

Total charge
= 2/3 − 1/3 − 1/3 = 0
Baryon number
= 1/3 + 1/3 + 1/3 = 1

Neutron = **udd**

Antiprotons are $\bar{u}\bar{u}\bar{d}$ and antineutrons are $\bar{u}\bar{d}\bar{d}$ — so no surprises there then.

There's no Such Thing as a Free Quark

What if you **blasted** a **proton** with **enough energy** — could you **separate out** the quarks?
Nope. The energy just gets changed into more **quarks and antiquarks** — it's **pair production** again and you just make **quark-antiquark pairs**. This is called **quark confinement**.

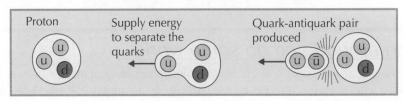

Proton   Supply energy to separate the quarks   Quark-antiquark pair produced

Quarks

Gluons Provide Force Between Quarks

1) When two particles **interact**, something must **happen** to let one particle know that the other one's there. That's the idea behind **exchange particles**. You can picture them if you think about **balls** and **boomerangs**:

Repulsion — Each time the **ball** is **thrown or caught** the people get **pushed apart**. It happens because the ball carries **momentum**.

Attraction — Each time the **boomerang** is **thrown or caught** the people **get pushed together**. (In real life, you'd probably fall in first.)

The particles don't <u>actually</u> loop round like that, though.

←— REPULSION —→   →ATTRACTION←

These exchange particles are called **gauge bosons** — they're virtual particles that only last for a very short time.

2) **All forces in nature** are caused by four **fundamental** forces. Each one has its **own gauge boson**:

Particle physicists never bother about gravity because it's so incredibly feeble compared with the other types of interaction. Gravity only really matters when you've got big masses like stars and planets. The graviton may exist but there's no evidence for it.

Type of Interaction	Gauge Boson	Particles Affected
strong	gluon	hadrons only
electromagnetic	photon	charged particles only
weak	W^+, W^-, Z^0	all types
gravity	graviton?	all types

3) The exchange particle that causes the **strong force** that 'glues' hadrons like protons and neutrons together is imaginatively called the **gluon**.

4) Because they cause a force, you can think of them as **fields** as well as particles. It's just the same as thinking of a **gravitational force** as being caused by a **gravitational field**.

5) As you try to **separate** quarks, you actually **increase** the **energy** of the gluon field, **increasing** the **attraction** between them.

6) If you keep pulling, eventually the energy in the gluon field will be enough that it produces a **quark-antiquark pair**. This is why you can **never** detect a quark on its own.

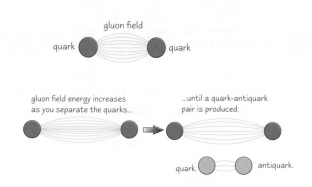

gluon field

quark   quark

gluon field energy increases as you separate the quarks...   ...until a quark-antiquark pair is produced.

quark   antiquark

Practice Questions

Q1 What is a quark?

Q2 Name the exchange particle for the strong force felt between two quarks.

Q3 Explain why quarks are never observed on their own.

Exam Questions

Q1 State the combination of three quarks that make up a neutron. [1 mark]

Q2 Give the quark composition of the proton. Explain how the charges of the quarks give rise to its charge. [2 marks]

A physical property called strangeness — how cool is that...

True, there's a lot of information here, but this page really does tie up a lot of the stuff on the last few pages. Learn as much as you can from this double page, then go back to page 60, and work back through to here. Don't expect to understand it all — but you will definitely find it much easier to learn when you can see how all the bits fit in together.

Particle Accelerators

Particle accelerators are devices that (surprisingly) accelerate particles, using electric and magnetic fields.
Accelerated particles can be used to investigate the fundamental particles that make up matter...

Particle Accelerators Cause High-Energy Collisions

There are lots of different types of accelerator out there smashing particles together.
One of the main types is the linear accelerator...

1) A **linear accelerator** is a long **straight** tube containing a series of **electrodes**.

2) **Alternating current** is applied to the electrodes so that their **charge** continuously **changes** between + and –.

3) The alternating current is **timed** so that the particles are always **attracted** to the **next electrode** in the accelerator and **repelled** from the **previous** one.

4) A particle's **speed** will **increase** each time it **passes** an electrode — so if the accelerator is long enough particles can be made to approach the **speed of light**.

5) The **high-energy particles** leaving a linear accelerator **collide** with a **fixed target** at the end of the tube.

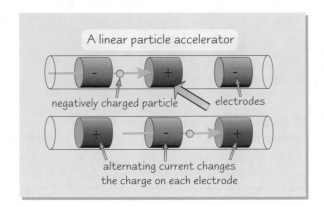

A linear particle accelerator

negatively charged particle   electrodes

alternating current changes
the charge on each electrode

A Cyclotron is a Circular Particle Accelerator

1) A cyclotron uses **two semicircular electrodes** to accelerate protons or other charged particles across a gap.

2) An **alternating potential difference** is applied between the electrodes — as the **particles** are **attracted** from one side to the other their **energy increases** (i.e. they are **accelerated**).

3) A **magnetic field** is used to keep the particles moving in a **circular motion** (in the diagram on the right, the magnetic field would be perpendicular to the page).

4) The combination of the **electric** and **magnetic fields** makes the particles **spiral outwards** as their energy increases.

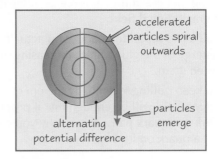

accelerated
particles spiral
outwards

particles
emerge

alternating
potential difference

Synchrotrons Produce Very High Energy Beams

1) A **synchrotron** can produce particle collisions with much **higher energies** than either a linear accelerator or a cyclotron.

2) **Electromagnets** keep the particles moving in a **circular path** in **focused beams**.

3) In this way, **synchrotrons** can produce particles with energies reaching from **500 GeV to several TeV**.

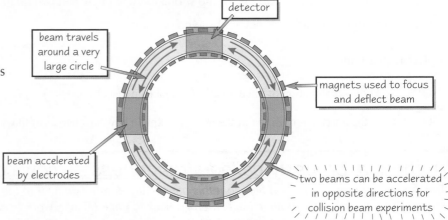

detector

beam travels
around a very
large circle

magnets used to focus
and deflect beam

beam accelerated
by electrodes

two beams can be accelerated
in opposite directions for
collision beam experiments

Particle Accelerators

Masses **Cannot** Reach the Speed of Light

1) Particles can be accelerated to such high speeds that the effects of **relativity** become noticeable and important.
2) Einstein's theory of **special relativity** only works in **reference frames** that **aren't accelerating** — called **inertial frames**.
3) It's based on two assumptions: **Physical laws have the same form in all inertial frames.**
   The speed of light in free space is invariant.
4) One rather interesting consequence of special relativity is that:

> No particle that has a mass can move at a speed greater than or equal to the speed of light, **c**.

The **Mass** of an Object **Increases** with **Speed**

1) Special relativity is where the idea that **energy is equivalent to mass** comes from.
2) It means that the more you **increase** the **kinetic energy** of a mass (like a particle in an accelerator), the **more massive** it gets.
3) This happens to any object with kinetic energy, but it's only noticeable at speeds approaching c.
4) Particle accelerators have to **alter** their magnetic and electric fields to compensate for the relativistic mass of the accelerating particles.

$$E = mc^2$$

The **Relativistic Factor** — $E_{tot} \div E_{rest}$

The **relativistic factor**, γ, (see page 33) is the total energy of a particle divided by the rest energy of the particle. For particles travelling at low speeds ($v \ll c$), γ will be very close to 1.

$$\gamma = \frac{E_{tot}}{E_{rest}}$$

$$E_{rest} = mc^2$$

Example

An electron is accelerated to almost the speed of light. It has a rest energy of 5.1×10^5 eV. If $\gamma = 235$, find the total energy of the accelerated electron in MeV.

$$E_{tot} = \gamma \times E_{rest} = 235 \times 5.1 \times 10^5$$
$$= 1.2 \times 10^8 \, eV$$
$$= 120 \, MeV$$

Practice Questions

Q1 Write down an expression for the relativistic factor.

Q2 Explain what happens when the energy of a particle travelling close to the speed of light increases.

Exam Question

Q1 A proton is accelerated by a synchrotron to a total energy of 500 GeV.
Show that the relativistic factor for a proton of this energy is about 500.
(The mass of a proton $m_p = 1.7 \times 10^{-27}$ kg.)   [4 marks]

Smash high-energy particles together to see what they're made of...

So, the three types of particle accelerator all have their advantages, but the synchrotron wins hands down on making very high energy particles for investigating fundamental particles. A famous synchrotron (in the physics world) is the Large Hadron Collider (LHC) found at CERN on the Swiss-French border — this accelerator is a whopping 27 km loop...

Electron Energy Levels

Electrons only exist in set energy levels. They leap into higher energy levels when they get excited.

Electrons in Atoms Exist in Discrete Energy Levels

1) **Electrons** in an **atom** can **only exist** in certain **well-defined energy levels**. Each level is given a **number** (called the **principal quantum number** of the electron in that state), with **n = 1** representing the electron's lowest possible energy — its **ground state**.

2) Electrons can **move down** an energy level by **emitting** a **photon**.

3) Since these **transitions** are between **definite energy levels**, the **energy** of **each photon** emitted can **only** take **certain values**.

4) The diagram on the right shows the **energy levels** for **atomic hydrogen**.

5) The **energies involved** are **so tiny** that it makes sense to use a more **appropriate unit** than the **joule**. The **electron-volt (eV)** is used instead:

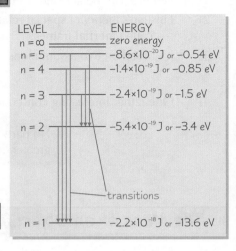

LEVEL   ENERGY
$n = \infty$ — zero energy
$n = 5$ — -8.6×10^{-20} J or -0.54 eV
$n = 4$ — -1.4×10^{-19} J or -0.85 eV
$n = 3$ — -2.4×10^{-19} J or -1.5 eV
$n = 2$ — -5.4×10^{-19} J or -3.4 eV
transitions
$n = 1$ — -2.2×10^{-18} J or -13.6 eV

> An **electron-volt** is the **kinetic energy carried** by an **electron** after it has been **accelerated** through a **potential difference** of **1 volt**.

$$1 \text{ eV} = 1.6 \times 10^{-19} \text{ J}$$

On the diagram, energies are labelled in **both units** for **comparison**.

6) All the electron energies are negative because of the way the zero energy is defined. All electrons that are **bound** to the atom have **negative** energies. The higher the energy level, the more energy the electron has and the less negative the energy. An electron is 'free' and no longer bound to the atom when it has a **potential energy** of **zero** — the atom becomes **ionised**.

7) The **energy** carried by each **photon** is **equal** to the **difference in energies** between the **two levels**. The following equation is for a **transition** between levels **n = 2** and **n = 1**: (where **h** = Planck's constant, **f** = frequency, **c** = speed of light in a vacuum, and **λ** = wavelength... but you probably knew all that from AS anyway.)

$$\Delta E = E_2 - E_1 = hf = \frac{hc}{\lambda}$$

8) In the same way, atoms can only **absorb** allowed photon energies. This **quantisation** of electron energies in atoms produces **line emission** and **absorption spectra** (see below).

> Electrons (as well as protons and neutrons) are **fermions**. That means they obey the **Pauli exclusion principle**. This states that **no two fermions** can be in **exactly** the same **quantum state** at the same time. In the context of energy levels, that means **no more than two** electrons can be in the same **energy level** at the same time.

The Evidence — Line Spectra

1) The **spectrum** of **white light** is **continuous**.

2) If you **split** the **light** up with a **prism**, the **colours** all **merge** into each other — there **aren't** any **gaps** in the spectrum.

Decreasing wavelength $\Longrightarrow$

3) You get a **line absorption spectrum** when **light** with a **continuous spectrum** passes through a **cool gas**.

4) At **low temperatures**, **most** of the **electrons** in the **gas atoms** will be in their **ground states**.

5) **Photons** of the **correct wavelength** are **absorbed** by the **electrons** to **excite** them to **higher energy levels**.

6) These **wavelengths** are then **missing** from the **continuous spectrum** when it **comes out** the other side of the gas.

7) You see a **continuous spectrum** with **dark lines** in it corresponding to the **absorbed wavelengths**.

8) When an electron falls into a **lower** energy level, it **emits** a photon. **Emission spectra** show the wavelengths of photons emitted. They are made up of a series **bright lines** corresponding to the **wavelengths emitted**.

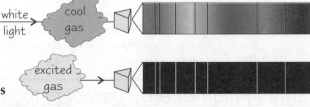

white light → cool gas
excited gas

9) If you **compare** the **absorption** and **emission** spectra of a **particular gas**, the **dark lines** in the **absorption spectrum** match up to the **bright lines** in the **emission spectrum**.

Electron Energy Levels

The **Wave Model** of the **Atom** can Help you Understand **Energy Levels**

1) Since light has both **particle** and **wave** characteristics, de Broglie suggested that **electrons** should have a **wave-like character**.

2) Specifically, when they're in orbit **around a nucleus** they ought to behave like the **standing waves** that are formed on a guitar string when it's plucked.

3) Just as standing waves on the guitar string only exist at certain **well-defined frequencies**, only certain standing waves are possible in an atom.

4) The **wavelength** of the electron waves should fit the **circumference** of the orbit a **whole number** of times.

5) The **principal quantum number** (corresponding to the number of the energy level) is equal to the number of **complete waves** that fit the circumference.

Three wavelengths
n = 3

Six wavelengths
n = 6

Not a standing wave
Forbidden energy

6) You can think of the electrons as being trapped by a **potential well** made by the nucleus. That way you can think of them as being standing waves between **two fixed walls**... so it's even more like a guitar string.

7) Erwin Schrödinger used the standing wave model and found that the **energy levels** in a **hydrogen atom** are given by:

$$E_n = \frac{-13.6 \text{ eV}}{n^2}$$

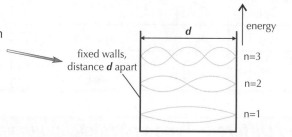

fixed walls,
distance **d** apart

energy

n=3
n=2
n=1

Practice Questions

Q1 Write down the equation you would use to find the difference in energy between two energy levels in an atom.

Q2 Describe the standing-wave model of electrons in an atom.

Q3 Describe how line spectra show the existence of discrete energy levels in atoms.

Exam Questions

Q1 The Balmer series is a series of spectral lines emitted by excited hydrogen atoms.
One Balmer line is caused by photons with a frequency of 4.57×10^{14} Hz.

(a) Find the energy of the photons that make up this line ($h = 6.6 \times 10^{-34}$ Js). [2 marks]

(b) The diagram shows some of the energy levels in a hydrogen atom.
Draw an arrow to show the energy level transition that causes this spectral line. [1 mark]

LEVEL	ENERGY
n = ∞	zero energy
n = 5	-8.6×10^{-20} J
n = 4	-1.4×10^{-19} J
n = 3	-2.4×10^{-19} J
n = 2	-5.4×10^{-19} J

Q2 The second is defined using the radiation from a particular quantum jump in the caesium-133 atom.
The difference in energy levels is 3.8×10^{-5} eV.

n = 1	-2.2×10^{-18} J

(a) Calculate the frequency of this radiation ($h = 6.6 \times 10^{-34}$ Js). [2 marks]

(b) State the number of oscillations that occur in 1 second, as used in the definition of the second. [1 mark]

My energy level's about n = 1 after that — I need chocolate...

I always find this the trickiest stuff to get — I mean, it's not like you can see an electron skipping about inside an atom every day, and the whole quarks and leptons stuff sounds like the cast list for a cheap sci-fi film. Once you've been through it all a few times though it does start to click, so stick with it and soon it'll be as easy as an electron transition from n = 2 to n = 1.

Radioactive Emissions

Congratulations, you've made it to the last section of the book — it's a good one, though, so don't skip over it.

Radioactive Decay is a Random Process

You should remember that **radioactive decay** is a **random process** that can be modelled by **exponential decay** (check out pages 4-5 if you don't know what I'm talking about). But that's not the whole story — there are **different kinds** of **nuclear radiation**, which all have different **uses** and different **risks**, and that's what this section is all about.

There are Four Types of Nuclear Radiation

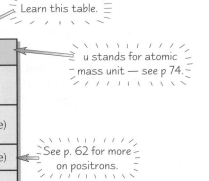

Learn this table.

u stands for atomic mass unit — see p 74.

Radiation	Symbol	Constituent	Relative Charge	Mass (u)
Alpha	α	A helium nucleus — 2 protons & 2 neutrons	+2	4
Beta-minus (Beta)	β or β⁻	Electron	−1	(negligible)
Beta-plus	β⁺	Positron	+1	(negligible)
Gamma	γ	Short-wave, high-frequency electromagnetic wave.	0	0

See p. 62 for more on positrons.

The Different Types of Radiation have Different Penetrations

Alpha, **beta** and **gamma** radiation can be **fired** at a **variety of objects** with **detectors** placed the **other side** to see whether they **penetrate** the object — as shown.

When a radioactive particle **hits** an atom it can **knock off electrons**, creating an **ion** — so, **radioactive emissions** are also known as **ionising radiation**.

The **table** shows the different **penetrating** and **ionising** properties.

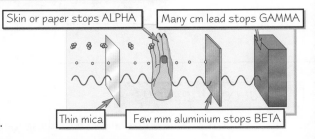

Skin or paper stops ALPHA   Many cm lead stops GAMMA

Thin mica   Few mm aluminium stops BETA

Radiation	Symbol	Ionising	Speed	Penetrating power	Affected by magnetic field
Alpha	α	Strongly	Slow	Absorbed by paper or a few cm of air	Yes
Beta-minus (Beta)	β or β⁻	Weakly	Fast	Absorbed by ~3 mm of aluminium	Yes
Beta-plus	β⁺	Annihilated by electron — so virtually zero range			
Gamma	γ	Very weakly	Speed of light	Absorbed by many cm of lead, or several m of concrete.	No

The Intensity of Gamma Radiation Decreases with Distance

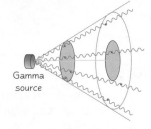

Gamma source

1) A **gamma source** will **emit** gamma **radiation** in **all directions**.

2) This radiation **spreads out** as you get **further away** from the source.

3) However, the amount of **radiation per unit area** (the **intensity**) will **decrease** the further you get from the source.

4) When gamma radiation travels through an **absorbing material** (e.g. concrete), its **intensity decreases exponentially**.

Radioactive Emissions

Radiation can Cause a lot of Harm to Body Tissues

1) The amount of **energy absorbed per kilogram** of tissue is called the **absorbed dose**, and is measured in **grays** (**Gy**).

$$\text{absorbed dose} = \frac{\text{energy}}{\text{mass}}$$

2) But the amount of **tissue damage** isn't just due to the amount of energy absorbed — it also depends on the **type of ionising radiation** and the **type of body tissue**.

3) The **effective dose** is a measure that lets you **compare the amount of damage** to body tissues that have been **exposed** to different types of radiation: **Effective dose = absorbed dose × radiation quality factor**

4) The **unit** of effective dose is the **sievert** (**Sv**).

5) The table shows typical values for the **radiation quality factor**. **For example**, if you exposed a sample of **body tissue** to **1 Gy** of **alpha** radiation, it could do the **same damage** as an exposure of 20 Gy of **gamma** radiation on the same type of body tissue.

Radiation	Typical radiation quality factor	Effective dose of 1 Gy
alpha	20	20 Sv
beta	1	1 Sv
gamma	1	1 Sv

Alpha and Beta Particles have Different Ionising Properties

The **different radiation quality factors** for each type of radiation can be **explained** by their **ionising properties**.

1) **Alpha** particles are **strongly positive** — so they can **easily pull electrons** off atoms, ionising them.

2) Ionising an atom **transfers** some of the **energy** from the **alpha particle** to the **atom**. The alpha particle **quickly ionises** many atoms (about 10 000 ionisations per alpha particle) and **loses** all its **energy** — that's why it causes so much **damage** to body tissue.

3) The **beta**-minus particle has **lower mass** and **charge** than the alpha particle, but a **higher speed**. This means it can still **knock electrons** off atoms. Each **beta** particle will ionise about 100 atoms, **losing energy** at each interaction.

4) This **lower** number of **interactions** means that beta radiation causes much **less damage** to body tissue than alpha radiation — explaining the **lower radiation quality factor**.

Risk = Probability × Consequences

Radioactive materials can be useful — for example, they're used to **generate power** (see pages 76-77), in **medicine** for **diagnosis** and **treatment,** and to **kill harmful microorganisms** that might contaminate our **food**. But they're also dangerous. They can cause **cancerous tumours**, **skin burns**, **sterility**, **radiation sickness**, **hair loss** and even **death**.

The result is that radiation is only used when the benefits outweigh the **risks**. There are two parts to the risk: **how likely** it is that the radiation will cause a problem, and **how bad** the problem would be if it did happen, e.g:

1) A **nuclear reactor** melting down would be **catastrophic**, but it's also **very unlikely**, so the risk might be acceptable.

2) Ionising radiation can **cause** cancer, but it can also be used in cancer **treatments** to destroy tumours. The risk of serious damage caused by the treatment is considered **acceptable** if the treatment is likely to **prolong** or **improve** a patient's life.

Practice Questions

Q1 Name three types of nuclear radiation and give three properties of each.

Q2 What does the term 'absorbed dose' mean?

Q3 What does the term 'effective dose' mean?

Exam Question

Q1 A mixed radiation source emits alpha, beta and gamma radiation.

(a) Draw a diagram showing the relative penetration of alpha, beta and gamma radiation. [3 marks]

(b) Safety precautions have to be taken when dealing with the source to prevent damage to the body tissues of the handler. Which of the following doses would cause the greatest damage to a sample of body tissue: 0.6 Gy of alpha radiation or 9 Gy of beta radiation? Show your working. [3 marks]

Radioactive emissions — as easy as α, β, γ...

It's a risky business using radiation, but remember that as well as the scary risk-of-death side of radioactive sources, they can do a lot of good too. Learn the different types of radiation and their properties... then go have a brew and a biccie...

Nuclear Decay

The stuff on these pages covers the most important facts about nuclear decay that you're just going to have to make sure you know inside out. I'd be very surprised if you didn't get a question about it in your exam...

Atomic Structure can be Represented Using Standard Notation

STANDARD NOTATION:

The nucleon number or mass number (A) — there are a total of 12 protons and neutrons in a carbon-12 atom.

The proton number or atomic number (Z) — there are six protons in a carbon atom.

$$^{12}_{6}C$$

The symbol for the element carbon.

Atoms with the **same number of protons** but **different numbers of neutrons** are called **isotopes**. The following examples are all isotopes of carbon: $^{12}_{6}C$, $^{13}_{6}C$, $^{14}_{6}C$

Some Nuclei are More Stable than Others

The nucleus is under the **influence** of the **strong nuclear force** holding it **together** and the **electromagnetic force pushing** the **protons apart**. It's a very **delicate balance**, and it's easy for a nucleus to become **unstable**. You can get a stability graph by plotting **Z** (atomic number) against **N** (number of neutrons).

A nucleus will be **unstable** if it has:

1) **too many neutrons**
2) **too few neutrons**
3) **too many nucleons** altogether, i.e. it's **too heavy**
4) **too much energy**

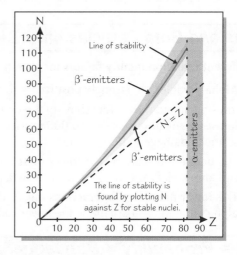

α *Emission Happens in Heavy Nuclei*

When an alpha particle is **emitted**:

nucleon number decreases by 4

The **proton number decreases** by **two**, and the **nucleon number decreases** by **four**.

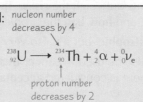

$$^{238}_{92}U \longrightarrow {}^{234}_{90}Th + {}^{4}_{2}\alpha + {}^{0}_{0}\nu_e$$

proton number decreases by 2

1) **Alpha emission** only happens in **very heavy** atoms (with more than 82 protons), like **uranium** and **radium**.

2) The **nuclei** of these atoms are **too massive** to be stable.

3) An alpha particle and a **neutrino** (ν_e) are emitted (see p60).

β^- *Emission Happens in Neutron Rich Nuclei*

1) **Beta-minus** (usually just called beta) decay is the emission of an **electron** from the **nucleus** along with an **antineutrino** (see p62).

2) Beta decay happens in isotopes that are **"neutron rich"** (i.e. have many more **neutrons** than **protons** in their nucleus).

3) When a nucleus ejects a beta particle, one of the **neutrons** in the nucleus is **changed** into a **proton**.

When a beta particle is **emitted**:

nucleon number stays the same

The **proton number increases** by **one**, and the **nucleon number stays the same**.

$$^{188}_{75}Re \longrightarrow {}^{188}_{76}Os + {}^{0}_{-1}\beta + {}^{0}_{0}\bar{\nu}_e$$

proton number increases by 1

In **beta-plus emission**, a **proton** gets **changed** into a **neutron**. The **proton number decreases** by **one**, and the **nucleon number stays the same**.

Nuclear Decay

γ *Radiation is Emitted from Nuclei with Too Much Energy*

1) A nucleus with **excess energy** is said to be **excited**.

2) This energy can be **lost** by emitting a **gamma ray**. This often happens after an **alpha** or **beta** decay.

> During **gamma emission**, there is **no change** to the nuclear **constituents** — the nucleus just **loses excess energy**.

There are Conservation Rules in Nuclear Reactions

In every nuclear reaction **energy, momentum, proton number / charge** and **nucleon number** must be conserved.

$$238 = 234 + 4 \text{ — nucleon numbers balance}$$

$$^{238}_{92}U \longrightarrow {}^{234}_{90}Th + {}^{4}_{2}\alpha$$

$$92 = 90 + 2 \text{ — proton numbers balance}$$

Support at the protest march was limited.

Mass is Not Conserved

1) The **mass** of the **alpha particle** is less than the **individual masses** of **two protons** and **two neutrons**. The difference is called the **mass defect**.

2) Mass **doesn't** have to be **conserved** because of **Einstein's equation**: $\boxed{E = mc^2}$

3) This says that **mass and energy** are **equivalent**.
   The **energy released** when the nucleons **bonded together** accounts for the missing mass — so the **energy released** is the same as the magnitude of the **mass defect × c^2**.

Practice Questions

Q1 What makes a nucleus unstable?

Q2 Describe the changes that happen in the nucleus during alpha, beta and gamma decay.

Q3 Explain the circumstances in which gamma radiation may be emitted.

Q4 Define the mass defect.

Exam Questions

Q1 Potassium-40 ($Z = 19$, $A = 40$) undergoes beta decay to calcium: ${}^{40}_{19}K \rightarrow {}^{40}_{20}Ca + {}^{0}_{-1}\beta + \overline{\nu}_e$
   Explain how you know from the equation that the antineutrino produced is uncharged. [1 mark]

Q2 (a) Radium-226 undergoes alpha decay to radon.
   Complete the equation for this reaction, showing nucleon and charge numbers.

   $${}^{226}_{88}Ra \rightarrow \quad Rn +$$
   [3 marks]

   (b) Calculate the energy released during the formation of an alpha particle, given that the total mass of two protons and two neutrons is 6.695×10^{-27} kg, the mass of an alpha particle is 6.645×10^{-27} kg and the speed of light, c, is 3.00×10^8 ms^{-1}. [3 marks]

Nuclear Decay — it can be enough to make you unstable...

$E = mc^2$ is an important equation that says mass and energy are equivalent. Remember it well, 'cos you're going to come across it a lot in questions about mass defect and the energy released in nuclear reactions over the next few pages...

Binding Energy

Turn off the radio and close the door, 'cos you're going to need to concentrate hard on this stuff about binding energy...

The **Mass Defect** is **Equivalent** to the **Binding Energy**

1) The **mass** of a **nucleus** is **less than** the mass of its **constituent parts**
   — this missing mass is called the **mass defect** (see previous page).

2) Einstein's equation, $E = mc^2$, says that mass and energy are **equivalent**.

3) So, as nucleons join together, the total mass **decreases** — this 'lost' mass is **converted** into energy and **released**.

4) The amount of **energy released** is **equivalent** to the **mass defect**.

5) If you **pulled** the nucleus completely **apart**, the **energy** you'd have to use to do it would be the **same** as the energy **released** when the nucleus formed.

> The energy needed to **separate** all of the nucleons in a nucleus is called the **binding energy** (measured in **MeV**), and it is **equivalent** to the **mass defect**.

> **Example** Calculate the binding energy of the nucleus of a lithium-6 atom, ^6_3Li, given that its mass defect is −0.0343 u.
>
>   1) Convert the mass defect into kg.
>
>   Mass defect = −0.0343 × 1.66 × 10⁻²⁷ = −5.70 × 10⁻²⁹ kg
>
>   2) Use $E = mc^2$ to calculate the binding energy.
>
>   E = −5.70 × 10⁻²⁹ × (3 × 10⁸)² = −5.13 × 10⁻¹² J = −32 MeV

Atomic mass is usually given in atomic mass units (u), where $1\ u = 1.66 \times 10^{-27}$ kg.

1 MeV = 1.6×10^{-13} J

6) The **binding energy per unit of mass defect** can be calculated (using the example above):

$$\frac{\text{Binding energy}}{\text{mass defect}} = \frac{-32\ \text{MeV}}{-0.0343\ \text{u}} \approx 931\ \text{MeV}\,\text{u}^{-1}$$

7) This means that a mass defect of **1 u ≈ 931 MeV** of binding energy.

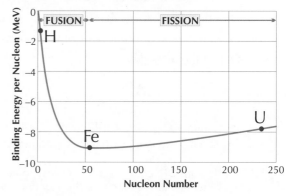

Captain Skip didn't believe in ghosts, marmalade and that things could be bound without rope.

The **Binding Energy Per Nucleon** is at a **Maximum** around **N = 50**

A useful way of **comparing** the binding energies of different nuclei is to look at the **binding energy per nucleon**.

> Binding energy per nucleon (in MeV) = $\dfrac{\text{Binding energy (B)}}{\text{Nucleon number (A)}}$

So, the binding energy per nucleon for ^6_3Li
(in the example above) is −32 ÷ 6 = −5.3 MeV.

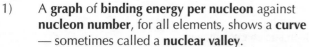

1) A **graph** of **binding energy per nucleon** against **nucleon number**, for all elements, shows a **curve** — sometimes called a **nuclear valley**.

2) The **most stable** nuclei occur around the **minimum point** on the graph — which is at **nucleon number 56** (i.e. **iron**, Fe). Nuclei with a nucleon number close to 56 are bound **most strongly**.

3) The more negative the binding energy per nucleon, the **more energy** is needed to **remove** nucleons from the nucleus.

4) **Combining small nuclei** is called nuclear **fusion** (see p. 77) — this **increases** the size of the **binding energy per nucleon** dramatically, which means a lot of **energy is released** during nuclear fusion.

5) **Fission** is when **large nuclei** are **split in two** (see p. 76) — the **nucleon numbers** of the two **new nuclei** are **smaller** than that of the original nucleus, which means the size of the **binding energy** per nucleon **increases**. So, energy is also **released** during nuclear fission (but not as much energy per nucleon as in nuclear fusion).

Binding Energy

The Change in Binding Energy Gives the Energy Released...

The **binding energy per nucleon graph** can be used to **estimate** the **energy released** from nuclear reactions.

Energy released in nuclear fusion

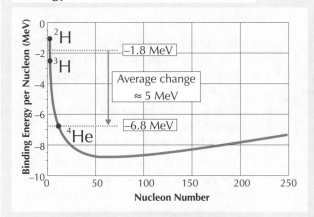

1) If **²H** and **³H** nuclei were **fused** together to form **⁴He**, the **average change** in binding energy per **⁴He** nucleon would be about **5 MeV**.

2) There are **4 nucleons** in **⁴He**, so we can **estimate** the **energy released** as $4 \times 5 = $ **20 MeV**.

Energy released in nuclear fission

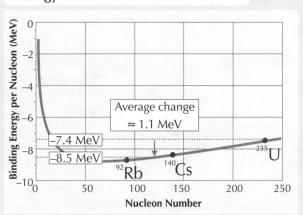

1) If a **²³⁵U** nucleus **splits** into **⁹²Rb** and **¹⁴⁰Cs** (plus a few neutrons) during nuclear **fission**, the **average change** in **binding energy per nucleon** would be about 1.1 MeV.

2) There are **235 nucleons** in **²³⁵U** to begin with, so we can **estimate** the energy **released** as $235 \times 1.1 \approx$ **260 MeV**.

Practice Questions

Q1 What is the binding energy of a nucleus?

Q2 How can you calculate the binding energy for a particular nucleus?

Q3 What is the binding energy per nucleon?

Q4 Which element has the most negative binding energy per nucleon?

Q5 Do nuclear fusion or fission reactions release more energy per nucleon?

Exam Questions

Q1 The mass of a $^{14}_{6}$C nucleus is 13.999948 u. The mass of a proton is 1.007276 u, and a neutron is 1.008665 u.
(a) Show that the mass defect of a $^{14}_{6}$C nucleus is -1.88×10^{-28} kg (given that 1 u = 1.66×10^{-27} kg). [3 marks]
(b) Use $E = mc^2$ to calculate the binding energy of the nucleus in MeV
(given that $c = 3 \times 10^8$ ms^{-1} and 1 MeV = 1.6×10^{-13} J). [2 marks]

Q2 The following equation represents a nuclear reaction that takes place in the Sun:

$$^1_1p + {}^1_1p \rightarrow {}^2_1H + {}^0_{+1}\beta + \text{energy released}$$
where p is a proton and β is a positron (see p. 62)

(a) State the name of this type of nuclear reaction. [1 mark]
(b) The binding energy per nucleon for a proton is 0 MeV and for a ²H nucleus it is approximately −0.86 MeV. Use this information to estimate the energy released by this reaction. [2 marks]

A mass of 1 u is equivalent to an energy of 931 MeV...

Remember this useful little fact, and it'll save loads of time in the exam — because you won't have to fiddle around with converting atomic mass from u → kg and binding energy from J → MeV. What more could you possibly want...

Nuclear Fission and Fusion

What did the nuclear scientist have for his tea? Fission chips... hohoho.

Fission *Means* Splitting Up *into* Smaller Parts

1) **Large nuclei**, with at least 83 protons (e.g. uranium), are **unstable** and some can randomly **split** into two **smaller** nuclei — this is called **nuclear fission**.

2) This process is called **spontaneous** if it just happens **by itself**, or **induced** if we **encourage** it to happen.

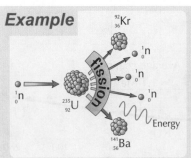

Example

Fission can be induced by making a neutron enter a ^{235}U nucleus, causing it to become very unstable.

Only low energy neutrons can be captured in this way. A low energy neutron is called a **thermal neutron**.

3) **Energy is released** during nuclear fission because the new, smaller nuclei have a **larger binding energy per nucleon** (see p. 74).

4) In general, the **larger** the nucleus, the more **unstable** it will be — so large nuclei are **more likely** to **spontaneously fission**.

5) This means that spontaneous fission **limits** the **number of nucleons** that a nucleus can contain — in other words, it **limits** the number of **possible elements**.

Controlled *Nuclear Reactors* Produce Useful *Power*

We can **harness** the **energy** released during nuclear **fission reactions** in a **nuclear reactor**, but it's important that these reactions are very **carefully controlled**.

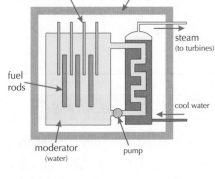

1) Nuclear reactors use **rods of uranium** that are rich in ^{235}U as 'fuel' for fission reactions. (The rods also contain a lot of ^{238}U, but that doesn't undergo fission.)

2) These **fission** reactions produce more **neutrons** which then **induce** other nuclei to fission — this is called a **chain reaction**.

3) The **neutrons** will only cause a chain reaction if they are **slowed down**, which allows them to be **captured** by the uranium nuclei — these slowed down neutrons are called **thermal neutrons**.

4) ^{235}U **fuel rods** need to be placed in a **moderator** (for example, **water**) to **slow down** and/or absorb **neutrons**. You need to choose a moderator that will slow down some neutrons enough so they can cause **further fission**, keeping the reaction going at a steady rate. Choosing a moderator that absorbs **more neutrons the higher the temperature** will **decrease** the chance of **meltdown** if the reactor **overheats** — as it will naturally **slow down** the reaction.

5) You want the chain reaction to continue on its own at a **steady rate**, where **one** fission follows another. The amount of 'fuel' you need to do this is called the **critical mass** — any less than the critical mass (**sub-critical mass**) and the reaction will just peter out. Nuclear reactors use a **supercritical** mass of fuel (where several new fissions normally follow each fission) and **control the rate of fission** using **control rods**.

6) Control rods control the **chain reaction** by **limiting** the number of **neutrons** in the reactor. They **absorb neutrons** so that the **rate of fission** is controlled. **Control rods** are made up of a material that **absorbs neutrons** (e.g. boron), and they can be inserted by varying amounts to control the reaction rate.
In an **emergency**, the reactor will be **shut down** automatically by the **release of the control rods** into the reactor, which will stop the reaction as quickly as possible.

7) **Coolant** is sent around the reactor to **remove heat** produced in the fission — often the coolant is the **same water** that is being used in the reactor as a **moderator**. The **heat** from the reactor can then be used to make **steam** for powering **electricity-generating turbines**.

If the chain reaction in a nuclear reactor is **left to continue unchecked**, large amounts of **energy** are **released** in a very **short time**.

Many new fissions will follow each fission, causing a **runaway reaction** which could lead to an **explosion**. This is what happens in a **fission (atomic) bomb**.

Nuclear Fission and Fusion

Waste Products of Fission Must be Disposed of Carefully

1) The **waste products** of **nuclear fission** usually have a **larger proportion of neutrons** than stable nuclei of a similar atomic number — this makes them **unstable** and **radioactive**.

2) The products can be used for **practical applications** such as **tracers** in medical diagnosis.

3) However, they may be **highly radioactive** and so their **handling** and **disposal** needs **great care**.

4) When material is removed from the reactor, it is initially **very hot**, so is placed in **cooling ponds** until the **temperature falls** to a safe level.

5) The radioactive waste is then **stored** underground in **sealed containers** until its **activity has fallen** sufficiently.

Fusion Means Joining Nuclei Together

1) **Two light nuclei** can **combine** to create a larger nucleus — this is called **nuclear fusion**.

2) Nuclei can **only fuse** if they have enough energy to overcome the **electrostatic repulsive** force between them, and get close enough for the **strong interaction** to bind them.

3) Typically they need about **1 MeV** of kinetic energy — and that's **a lot of energy**.

Example

In the Sun, **hydrogen nuclei** fuse in a series of reactions to form **helium**.

$$^2_1H + {}^1_1H \rightarrow {}^3_2He + \text{energy}$$

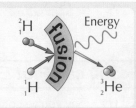

Fusion Happens in the Core of Stars

1) The **energy** emitted by the **Sun** and other stars comes from nuclear **fusion** reactions.

2) Fusion can happen because the **temperature** in the **core of stars** is so **high** — the core of the Sun is about 10^7 K.

3) At these temperatures, **atoms don't exist** — the negatively charged electrons are **stripped away**, leaving **positively charged nuclei** and **free electrons**. The resulting mixture is called a **plasma**.

4) A lot of **energy** is released during nuclear fusion because the new, heavier nuclei have a **much larger binding energy per nucleon** (see p. 74). This helps to **maintain the temperature** for further fusion reactions to happen.

5) Experimental **fusion reactors** like JET (the Joint European Torus) are trying to **recreate** these conditions to generate **electricity** (without all the nasty waste you get from fission reactors). Unfortunately, the electricity generated at the moment is **less** than the amount needed to get the reactor up to temperature. But watch this space...

Practice Questions

Q1 What is spontaneous fission?

Q2 How can fission be induced in ^{235}U?

Q3 Why must the waste products of nuclear fission be disposed of very carefully?

Q4 Describe the conditions in the core of a star.

Exam Questions

Q1 Nuclear reactors use carefully controlled chain reactions to produce energy.
   (a) Explain what is meant by the expression 'chain reaction' in terms of nuclear fission. [2 marks]
   (b) Describe and explain one feature of a nuclear reactor whose role is to control the rate of fission.
       Include an example of a suitable material for the feature you have chosen. [3 marks]
   (c) Explain what happens in a nuclear reactor during an emergency shut-down. [2 marks]

Q2 Discuss two advantages and two disadvantages of using nuclear fission to produce electricity. [4 marks]

If anyone asks, I've gone fission... that joke never gets old...

So, controlled nuclear fission reactions can provide a shedload of energy to generate electricity. There are pros and cons to using fission reactors. They produce huge amounts of energy without so much greenhouse gas, but they leave behind some very nasty radioactive waste... But then, you already knew that — now you need to learn all the grisly details.

Exponentials and Natural Logs

Mwah ha ha ha... you've hacked your way through the rest of the book and think you've finally got to the end of A2 Physics, but no, there's this tasty titbit of exam fun to go. You can get asked to look at and work out values from log graphs all over the shop, from astrophysics to electric field strength. And it's easy when you know how...

Many Relationships in Physics are **Exponential**

A fair few of the relationships you need to know about in A2 Physics are **exponential** — where the **rate of change** of a quantity is **proportional** to the **amount** of the quantity left. Here are just a couple you should have met before (if they don't ring a bell, go have a quick read about them)...

Charge on a capacitor — the decay of charge on a capacitor is proportional to the amount of charge left on the capacitor:
$$Q = Q_o\, e^{(-t/RC)}$$
(see p. 9)

Radioactive decay — the rate of decay is proportional to the **number of nuclei left** to decay in a sample:
$$N = N_o\, e^{(-\lambda t)}$$
(see p. 5)

You can **Plot** Exponential Relations Using the **Natural Log, ln**

1) Say you've got two variables, *x* and *y*, which are related to each other by the formula $y = ke^{-ax}$ (where *k* and *a* are constants).

2) The inverse of e is the natural logarithm, **ln**.

3) By definition, $\ln(e^x) = x$. So far so good... now you need some **log rules**:

$$\ln(ab) = \ln a + \ln b \qquad \ln\left(\frac{a}{b}\right) = \ln a - \ln b \qquad \ln a^b = b\ln a$$

When it came to logs, Geoff always took time to smell the flowers...

~ You don't need to learn these log rules — you're given them on your formula sheet. ~

4) So, if you take the natural log of the exponential function you get:

$$\ln y = \ln(ke^{-ax}) = \ln k + \ln(e^{-ax}) \implies \boxed{\ln y = \ln k - ax}$$

5) Then all you need to do is plot (ln *y*) against *x*, and Eric's your aunty: $\implies$

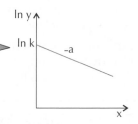

You get a **straight-line graph** with (**ln k**) as the **y-intercept**, and **–a** as the **gradient**.

You Might be Asked to find the **Gradient** of a Log Graph...

This log business isn't too bad when you get your head around which bit of the log graph means what. On the plus side, they won't ask you to plot a graph like this (yipee) — they'll just want you to find the **gradient** or the **y-intercept**.

Example — finding the radioactive half-life of material X

The graph shows the radioactive decay of substance X.
(a) Find the initial number of atoms, N_o, in the sample.

You know that the number of radioactive atoms in a sample, *N*, is related to the initial number of atoms by the equation $N = N_o e^{-\lambda t}$.

So, $(\ln N) = (\ln N_o) - \lambda t$ and $\ln N_o$ is the y-intercept of the graph = 9.2, $N_o = e^{9.2} \approx$ **9900 atoms**.

(b) Find the decay constant λ of substance X.

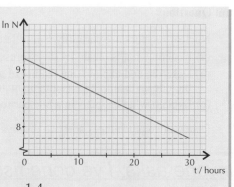

$-\lambda$ is the gradient of the graph, so: $\lambda = \dfrac{\Delta \ln N}{\Delta t} = \dfrac{9.2 - 7.8}{30 \times 60 \times 60} = \dfrac{1.4}{108\,000} = 1.3 \times 10^{-5}\ \text{s}^{-1}$ (2 s.f.)

Log Graphs and Long Answer Questions

You can Plot **Any Power Law** as a **Log-Log Graph**

You can use logs to plot a straight-line graph of **any power law** — it doesn't have to be an exponential.

Take the relationship between the energy stored in a spring, **E**, and the spring's extension, **x**:

$$E = kx^n$$

Take the log (base 10) of both sides to get:

$$\boxed{\log E = \log k + n \log x}$$

So **log k** will be the **y**-intercept and **n** the gradient of the graph.

Example

The graph shows how the intensity of radiation from the Sun, **I**, varies with its distance, **d**.
I is related to **d** by the power law **I = kd^n**. Find **n**.

$\log I = \log (kd^n) = \log k + \log d^n$
$= \log k + n \log d.$

so **n** is the **gradient** of the graph.
Reading from the graph:

$$n = \frac{\Delta \log I}{\Delta \log d} = \frac{15.4 - 5.4}{5 - 10} = \frac{10}{-5} = -2$$

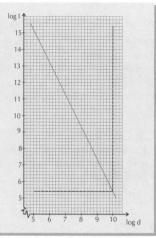

And that's the End of Logs... Now **Explain Yourself**...

In A2, they often give a couple of marks for 'the quality of written communication' when you're writing a slightly long answer (and not just pumping numbers into an equation).
You can pick up a couple of easy marks just by making sure that you do the things in the fetching blue box.

1) **Explain** your ideas or argument **clearly**, as this is usually what you'll get a mark for. And make sure you **answer the question** being asked — it's dead easy to go off on a tangent. Like my mate Phil always says... have I ever told you about Phil? Well he...

2) Write in **whole sentences**.

3) Use **correct spelling**, **grammar** and **punctuation**.

4) Also check how many marks the question is worth. If it's only a two-marker, they don't want you to spend half an hour writing an essay about it.

Example

A large group of people walk across a footbridge. When the frequency of the group's footsteps is 1 Hz, the bridge noticeably oscillates and 'wobbles'.
Fully describe the phenomenon causing the bridge to wobble.
Suggest what engineers could do to solve this problem.
The quality of your written answer will be assessed in this question.   [6 marks]

Good Answer

The pedestrians provide a driving force on the bridge, causing it to oscillate. At around 1 Hz, the driving frequency from the pedestrians is roughly equal to the natural frequency of the bridge, causing it to resonate. The amplitude of the bridge's oscillations when resonating at 1 Hz will be greater than at any other driving frequency. The oscillations at this frequency are large enough to be noticed by pedestrians.

Engineers could fix this problem by critically damping the bridge to stop any oscillations as quickly as possible.

They could also adjust the natural frequency of the bridge so that it was not so close to a known walking frequency of large groups of people.

Bad Answer

resonance
driving frequency of group = nat. freq.
damping

There's nothing wrong with the physics in the bad answer, but you'd miss out on some nice easy marks just for not bothering to link your thoughts together properly or put your answer into proper sentences.

Lumberjacks are great musicians — they have a natural logarithm...

Well, that's it folks. Crack open the chocolate bar of victory and know you've earnt it. Only the tiny detail of the actual exam to go... ahem. Make sure you know which bit means what on a log graph and you'll pick up some nice easy marks. Other than that, stay calm, be as clear as you can and good luck — I've got my fingers, toes and eyes crossed for you.

Answers

Unit 4: Section 1 — Creating Models
Page 5 — Radioactivity and Exponential Decay

1) a) $T_{\frac{1}{2}} = \dfrac{\ln 2}{\lambda} = \dfrac{0.693}{0.014 \times 10^{-3}} = 49500$ seconds

 [1 mark for the half-life equation, 1 mark for the correct half-life]

 b) $N = N_0 e^{-\lambda t} = 50000 \times e^{-0.000014 \times 300} = 49750$

 [2 marks available — 1 mark for the decay equation, 1 mark for the number of atoms remaining after 300 seconds]

 c) E.g. Models allow you to simplify a process to be able to make predictions *[1 mark for any sensible answer].*

Page 7 — Capacitors

1) Capacitors are suitable because they can deliver a short pulse of high current *[1 mark],* which results in a brief flash of bright light when needed *[1 mark].*

2) a) $E = \dfrac{1}{2}CV^2 = \dfrac{1}{2} \times 0.5 \times 12^2 = 36$ J *[1 mark]*

 b) $Q = CV$ *[1 mark]* $= 0.5 \times 12 = 6$ C *[1 mark]*

Page 9 — Charging and Discharging

1) a) The charge falls to 37% after RC seconds *[1 mark],*
 so $t = 1000 \times 2.5 \times 10^{-4} = 0.25$ seconds *[1 mark]*

 b) $Q = Q_0 e^{-\frac{t}{RC}}$ *[1 mark],* so after 0.7 seconds: $Q = Q_0 e^{-\frac{0.7}{0.25}} = Q_0 \times 0.06$

 [1 mark]. There is 6% of the initial charge left on the capacitor after 0.7 seconds *[1 mark].*

Page 11 — Modelling Decay

1) $RC = 500 \times 10^{-6} \times 144 \times 10^3 = 72$ *[1 mark]*
 $\lambda = 1 \div 72 = 0.0139 \ s^{-1}$ *[1 mark]*

2)

Time (s)	Charge (C)	dQ/dt (Cs⁻¹)	ΔQ (C)	New charge (C)
0	5.0×10^{-2}	-2.5×10^{-2}	-1.25×10^{-2}	3.75×10^{-2}
0.5	3.75×10^{-2}	-1.88×10^{-2}	-0.94×10^{-2}	2.81×10^{-2}
1.0	2.81×10^{-2}	-1.41×10^{-2}	-0.70×10^{-2}	2.11×10^{-2}
1.5	2.11×10^{-2}	-1.05×10^{-2}	-0.53×10^{-2}	1.58×10^{-2}
2.0	1.58×10^{-2}			

So the answer is $\mathbf{1.58 \times 10^{-2}}$ C

[1 mark for each correct stage of iteration, to a total of 5 marks. Otherwise 1 mark for an attempt at an iterative method]

Page 13 — Simple Harmonic Motion

1) a) Simple harmonic motion is an oscillation in which an object always accelerates towards a fixed point *[1 mark]* with an acceleration directly proportional to its displacement from that point *[1 mark].*
 [The SHM equation would get you the marks if you defined all the variables.]

 b) The acceleration of a falling bouncy ball is due to gravity. This acceleration is constant, so the motion is not SHM. *[1 mark]*

2) a) Maximum velocity $= (2\pi f)A = 2\pi \times 1.5 \times 0.05 = 0.47 \ ms^{-1}$ *[1 mark].*

 b) Stopclock started when object released, so $x = A\cos(2\pi ft)$ *[1 mark].*
 $x = 0.05 \times \cos(2\pi \times 1.5 \times 0.1) = 0.05 \times \cos(0.94) = 0.029$ m *[1 mark].*

Page 15 — Simple Harmonic Oscillators

1) a) Extension of spring $= 0.20 - 0.10 = 0.10$ m *[1 mark].* Hooke's Law gives $k = \dfrac{force}{extension}$, so $k = \dfrac{0.10 \times 9.8}{0.10} = 9.8 \ Nm^{-1}$ *[1 mark].*

 b) $m \propto T^2$ so if T is doubled, T^2 is quadrupled and m is quadrupled *[1 mark].*
 So mass needed $= 4 \times 0.10 = 0.40$ kg *[1 mark].*

2) $5T_{short \ pendulum} = 3T_{long \ pendulum}$, and $T = 2\pi\sqrt{\dfrac{l}{g}}$ *[1 mark].*

 Let length of long pendulum $= l$. So $5\left(2\pi\sqrt{\dfrac{0.20}{g}}\right) = 3\left(2\pi\sqrt{\dfrac{l}{g}}\right)$ *[1 mark].*

 Dividing by 2π gives $5 \times \sqrt{\dfrac{0.20}{g}} = 3 \times \sqrt{\dfrac{l}{g}}$. Squaring and simplifying

 gives $5 = 9l$ so length of long pendulum $= 5/9 = 0.56$ m *[1 mark].*

Page 17 — Free and Forced Vibrations

1) a) When a system is forced to vibrate at a frequency that's close to, or the same as, its natural frequency *[1 mark]* and oscillates with a much larger than usual amplitude *[1 mark].*

 b) See graph below. *[1 mark for showing a peak at the natural frequency, 1 mark for a sharp peak.]*

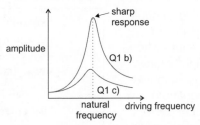

 c) See graph. *[1 mark for a smaller peak at the natural frequency]*
 (The peak will actually be slightly to the left of the natural frequency due to the damping, but you'll get the mark if the peak is at the same frequency in the diagram.)

2) It does not oscillate *[1 mark]* and returns to its equilibrium position in the shortest time possible *[1 mark].*

Unit 4: Section 2 — Out into Space
Page 19 — Forces and Momentum

1) total momentum before collision = total momentum after *[1 mark]*
 $(0.6 \times 5) + 0 = (0.6 \times -2.4) + 2v$
 $3 + 1.44 = 2v$ *[1 mark for working]* $\Rightarrow v = 2.22 \ ms^{-1}$ *[1 mark]*

2) momentum before = momentum after *[1 mark]*
 $(0.7 \times 0.3) + 0 = 1.1v$
 $0.21 = 1.1v$ *[1 mark]* $\Rightarrow v = 0.191 \approx 0.19 \ ms^{-1}$ *[1 mark]*

3) total momentum before engines started $= 0 \ kg \ ms^{-1}$
 so, total momentum after engines started must also $= 0 \ kg \ ms^{-1}$
 [1 mark] $(4200 \times 2.5) + p = 0 \Rightarrow p = -10 \ 500 \ kg \ ms^{-1}$ *[1 mark]*

Answers

Page 21 — Newton's Laws of Motion

1)a) $F = ma = 78 \times 9.81 = 765.18$ N [1 mark]

b) Newton's 1st law states that the velocity of an object will not change unless a resultant force is acting upon it [1 mark] — the parachutist's velocity is not changing, so the resultant force acting on her must be zero [1 mark].

2) Force perpendicular to river flow = 500 − 100 = 400 N [1 mark]
Force parallel to river flow = 300 N

Resultant force = $\sqrt{400^2 + 300^2}$ = 500 N [1 mark]

$a = F/m$ [1 mark] = 500/250 = 2 ms^{-2} [1 mark]

Page 23 — Work and Energy

1)

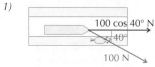

Force in direction of travel = 100 cos 40° = 76.6 N [1 mark]
$W = Fs$ = 76.6 × 1500 = 114 900 J ≈ 115 kJ [1 mark]

2)a) Use $W = Fs$ [1 mark] = 20 × 9.81 × 3 = 588.6 J [1 mark]

b) Use $P = Fv$ [1 mark] = 20 × 9.81 × 0.25 = 49.05 W [1 mark]

Page 25 — Circular Motion

1)a) $v = r\omega$ [1 mark] = $1.5 \times 10^{11} \times 2.0 \times 10^{-7}$ = 30 kms^{-1} [1 mark]

b)i) $F = m\omega^2 r$ [1 mark] = $6.0 \times 10^{24} \times (2.0 \times 10^{-7})^2 \times 1.5 \times 10^{11}$
= 3.6×10^{22} N [1 mark]

ii) The gravitational force between the Sun and the Earth [1 mark]

2)a) 9.81ms^{-2} (as that's the acceleration due to gravity). [1 mark]

b) Since $a = \omega^2 r$, $\omega^2 = \dfrac{a}{r} = \dfrac{9.81}{1}$, so ω = 3.1 $rad\,s^{-1}$ [1 mark]

$\omega = 2\pi f$, so $f = \dfrac{\omega}{2\pi}$ = 0.5 $rev\,s^{-1}$ [1 mark]

Page 27 — Gravitational Fields

1) $F = -\dfrac{GMm}{r^2} \Rightarrow M = -\dfrac{r^2}{G} \times \dfrac{F}{m}$ [1 mark] and $g = \dfrac{F}{m}$ so,

$M = -\dfrac{gr^2}{G}$ [1 mark] $\Rightarrow M = -\dfrac{(-9.81) \times (6400 \times 1000)^2}{6.67 \times 10^{-11}}$ [1 mark]

= 6.02×10^{24} kg [1 mark]

2)a) $F = -\dfrac{GMm}{r^2} = -\dfrac{6.67 \times 10^{-11} \times 7.35 \times 10^{22} \times 25}{(1740 \times 10^3)^2}$ = −40.5 N

[2 marks, otherwise 1 mark for some correct working]

b) $\Delta PE = mg\Delta h$ = 25 × 1.64 × 2 × 10^3 = 82 000 J
[2 marks, otherwise 1 mark for some correct working]

Page 29 — Gravitational Fields

1)a) $g = -\dfrac{GM}{r^2} = -\dfrac{6.67 \times 10^{-11} \times 5.98 \times 10^{24}}{(6600 \times 10^3)^2}$ = −9.16 $N\,kg^{-1}$

[2 marks, otherwise 1 mark for some correct working]

b) $E_{grav} = -\dfrac{GMm}{r^2} = -\dfrac{6.67 \times 10^{-11} \times 5.98 \times 10^{24} \times 3015}{6600 \times 10^3}$ = −1.9 × 10^{11} J

[2 marks, otherwise 1 mark for some correct working]

c) $v = \sqrt{\dfrac{GM}{r}} = \sqrt{\dfrac{6.67 \times 10^{-11} \times 5.98 \times 10^{24}}{6600 \times 10^3}}$ = 7774 ms^{-1} = 7.77 kms^{-1}

[2 marks, otherwise 1 mark for some correct working]

2) Over 50 000 years, the Sun will have only lost a tiny fraction of its mass (9.5 × 10^{21} kg overall) [1 mark], which will not have caused any significant change in the Earth's orbit [1 mark].

Unit 4: Section 3 — Our Place in the Universe
Page 31 — The Solar System & Astronomical Distances

1)a) $2d = ct$ [1 mark], so $d = [3 \times 10^8 \times (4.6 \times 60)] \div 2$ [1 mark]
d = (8.28 × 10^{10}) ÷ 2 = 4.14 × 10^{10} ≈ 4.1 × 10^{10} m [1 mark]

b) Any two of: The speed of the radio waves (the speed of light) is constant [1 mark]. The time taken for the radio waves to reach Venus is the same as the time taken for them to return to Earth [1 mark]. The speed of Venus relative to the Earth is much less than the speed of light [1 mark].

2)a) A light-year is the distance travelled by electromagnetic waves through a vacuum in one year [1 mark].

b) Seconds in a year = 365.25 × 24 × 60 × 60 = 3.16 × 10^7 s [1 mark].
Distance = c × time = 3.0 × 10^8 × 3.16 × 10^7 = 9.5 × 10^{15} m [1 mark].

c) To see something, light must reach us. Light travels at a finite speed, so it takes time for that to happen [1 mark]. The further out we see, the further back in time we're looking. The Universe is ~14 billion years old so we can't see further than ~14 billion light years [1 mark].

Page 33 — The Doppler Effect and Redshift

1)a) Object A is moving towards us [1 mark].

b) Find the velocity of object B using $\dfrac{\Delta\lambda}{\lambda} = \dfrac{v}{c}$, so $v = c\dfrac{\Delta\lambda}{\lambda}$. [1 mark]

$v = 3.0 \times 10^8 \times \dfrac{(667.83 \times 10^{-9} - 656.28 \times 10^{-9})}{656.28 \times 10^{-9}}$ = $5.28 \times 10^6\,ms^{-1}$
So object B is moving away from us [1 mark] at 5.28 × 10^6 ms^{-1} [1 mark].

Page 35 — The Big Bang Model of the Universe

1)a) Hubble's law suggests that the Universe originated with the Big Bang [1 mark] and has been expanding ever since. [1 mark]

b)i) $H_0 = v \div d$ = 50 kms^{-1} ÷ 1 Mpc.
50 kms^{-1} = 50 × 10^3 ms^{-1} and 1 Mpc = 3.09 × 10^{22} m
So, H_0 = 50 × 10^3 ms^{-1} ÷ 3.09 × 10^{22} m = 1.62 × 10^{-18} s^{-1}
[1 mark for the correct value, 1 mark for the correct unit]

ii) $t = 1/H_0$ [1 mark]
t = 1/1.62 × 10^{-18} = 6.18 × 10^{17} s ≈ 20 billion years [1 mark]
The observable Universe has a radius of 20 billion light years. [1 mark]

2)a) $z \approx v/c$ [1 mark] so $v \approx 0.37 \times 3.0 \times 10^8 \approx 1.1 \times 10^8\,ms^{-1}$ [1 mark]

b)i) $d = v/H_0 \approx 1.1 \times 10^8 / 2.4 \times 10^{-18}$ = 4.6 × 10^{25} m [1 mark]
= 4.6 × 10^{25} / 9.5 × 10^{15} ly = 4.9 billion ly [1 mark]

ii) $z = v/c$ is only valid if $v \ll c$ — it isn't in this case [1 mark].

3) It suggests that the ancient Universe was very hot, producing lots of electromagnetic radiation [1 mark] and that its expansion has stretched the radiation into the microwave region [1 mark].

Answers

Unit 4: Section 4 — Matter: Very Simple
Page 37 — Ideal Gases
1) The graph plotted will be straight line. It will cross the temperature axis at absolute zero (−273.15 K). [1 mark]

2) a) Number of moles = $\dfrac{\text{mass of gas}}{\text{molar mass}} = \dfrac{0.014}{0.028} = 0.5$ [1 mark]

   b) $pV = nRT$, so $p = \dfrac{nRT}{V}$ [1 mark] $= \dfrac{0.5 \times 8.31 \times 300}{0.01}$

   = 125 000 Pa [1 mark]

   c) The pressure would also halve because it is proportional to the number of molecules — $pV = NkT$ [1 mark].

3) At ground level, $\dfrac{pV}{T} = \dfrac{1 \times 10^5 \times 10}{293} = 3410\ JK^{-1}$ [1 mark]

   pV/T is constant, so higher up $pV/T = 3410\ JK^{-1}$ [1 mark]

   Higher up, $p = \dfrac{3410 \times T}{V} = \dfrac{3410 \times 260}{25} = 35\ 500$ Pa [1 mark]

Page 39 — The Pressure of an Ideal Gas
1) a) $pV = \dfrac{1}{3}Nm\overline{c^2}$ [1 mark] Rearrange the equation:

   $\overline{c^2} = \dfrac{3pV}{Nm} = \dfrac{3 \times 1 \times 10^5 \times 7 \times 10^{-5}}{2 \times 10^{22} \times 6.6 \times 10^{-27}} = 159\ 091\ (ms^{-1})^2$ [1 mark]

   b) r.m.s. speed $= \sqrt{\overline{c^2}} = \sqrt{159\ 091} = 399\ ms^{-1}$ [1 mark]

   c) The increase in temperature will increase the average speed and therefore momentum of the gas particles [1 mark]. This means that as the particles will hit the walls of the flask there will be a greater change in momentum, exerting a greater force on the walls of the flask [1 mark]. As the particles are travelling faster they will also hit the walls (and exert a force) more frequently [1 mark]. This leads to an increase in pressure (as pressure = force ÷ area).

Page 41 — Internal Energy and Temperature
1) a) Molecule mass $= \dfrac{\text{mass of 1 mole}}{N_A} = \dfrac{2.8 \times 10^{-2}}{6.02 \times 10^{23}}$

   = 4.65×10^{-26} kg [1 mark]

   b) $\dfrac{1}{2}m\overline{c^2} = \dfrac{3kT}{2}$ Rearranging gives: $\overline{c^2} = \dfrac{3kT}{m}$ [1 mark]

   $\overline{c^2} = \dfrac{3 \times 1.38 \times 10^{-23} \times 300}{4.65 \times 10^{-26}} = 2.67 \times 10^5\ m^2s^{-2}$ [1 mark]

   Typical speed = r.m.s. speed $= \sqrt{2.67 \times 10^5} = 516.7 \approx 520\ ms^{-1}$ [1 mark]

   c) Gas molecules move at different speeds because they have different amounts of energy [1 mark]. The molecules have different amounts of energy because they are constantly colliding and transferring energy between themselves [1 mark].

2) a) Speed = $\dfrac{\text{distance}}{\text{time}}$ so time = $\dfrac{\text{distance}}{\text{speed}} = \dfrac{8.0\ m}{400\ ms^{-1}} = 0.02$ s [1 mark]

   b) Although the particles are moving at an average of 400 ms^{-1}, they are frequently colliding with other particles. [1 mark]
   This means their motion in any one direction is limited and so they only slowly move from one end of the room to the other. [1 mark]

   c) At 30 °C the average speed of the particles would be slightly faster [1 mark] since the absolute temperature would have risen (from 293 K to 303 K) and the temperature determines the average speed [1 mark].
   This means the speed of diffusion would also be faster [1 mark].

Unit 4: Section 5 — Matter: Hot or Cold
Page 43 — Activation Energy
1) a) Average thermal energy is approximately kT.
   i) $300k = 300 \times 1.38 \times 10^{-23} = 4 \times 10^{-21}$ J [1 mark]
   ii) $360k = 360 \times 1.38 \times 10^{-23} = 5 \times 10^{-21}$ J [1 mark]

   b) An activation energy is needed to break the strong attractive forces between the particles in the oil [1 mark]. The higher average energy of the particles at 360 K means they are more likely to have an energy greater than the activation energy [1 mark].

Page 45 — The Boltzmann Factor
1) a) $kT = 1.38 \times 10^{-23} \times 300 \approx 4 \times 10^{-21}$ J [1 mark]

   b) For two bonds, $\varepsilon = 2 \times 3.2 \times 10^{-20} = 6.4 \times 10^{-20}$ J [1 mark]

   c) $\dfrac{\varepsilon}{kT} = \dfrac{6.4 \times 10^{-20}}{4 \times 10^{-21}} \approx 16$ [1 mark]

   d) With an ε/kT ratio of about 16, processes can take place using random thermal energy [1 mark]. Although the average energy of a water molecule is much less than it needs to escape [1 mark], some molecules will have enough energy to break their bonds and escape, meaning that the tank must be topped up to replace the water lost by evaporation [1 mark].

Unit 5: Section 1 — Electromagnetic Machines
Page 47 — Magnetic Fields and Motors
1) a) 0 N [1 mark] — the current is parallel to the flux lines so no force acts.
   b) $F = BIl$ [1 mark]
   $= 2 \times 10^{-5} \times 3 \times 0.04 = 2.4 \times 10^{-6}$ N [1 mark]

Page 49 — Electromagnetic Induction
1) a) $\Phi = BA$ [1 mark] $= 2 \times 10^{-3} \times 0.23 = 4.6 \times 10^{-4}$ Wb [1 mark]
   b) flux linkage = ΦN [1 mark] $= 4.6 \times 10^{-4} \times 150 = 0.069$ Wb [1 mark]

   c) $V = \dfrac{d(\Phi N)}{dt}$ [1 mark]

   $= \dfrac{(4.6 \times 10^{-4} - 3.5 \times 10^{-4}) \times 150}{2.5} = 6.6 \times 10^{-3}$ V [1 mark]

2) B, the movement will only induce a current if the rod is part of a complete circuit [1 mark].

3)
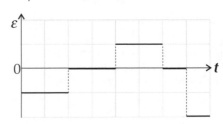

Step graph [1 mark], with the first and third steps negative and second step positive [1 mark] and the last step twice the height of the others [1 mark].

Page 51 — Transformers and Alternators

1) a) $\dfrac{V_p}{V_s} = \dfrac{N_p}{N_s}$ *[1 mark]* so, $N_s = \dfrac{45 \times 150}{9} = 750$ turns *[1 mark]*

   b) $\dfrac{V_p}{V_s} = \dfrac{I_s}{I_p}$ *[1 mark]* so, $I_s = \dfrac{V_p I_p}{V_s} = \dfrac{9 \times 1.5}{45} = 0.3$ A *[1 mark]*

   c) efficiency $= \dfrac{V_s I_s}{V_p I_p}$ *[1 mark]* $= \dfrac{10.8}{9 \times 1.5} = 0.8$ (i.e. 80%) *[1 mark]*

2) AC current flowing in the primary coil creates a changing magnetic flux/field in the core of the transformer *[1 mark]*. This induces an e.m.f. and current in the core *[1 mark]*, which creates a magnetic flux/field in the core that opposes the original change of flux *[1 mark]*. *[1 mark for a clear sequence of ideas.]*

Unit 5: Section 2 — Charge and Field
Page 53 — Electric Fields

1)

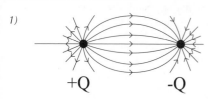

$+Q \qquad\qquad -Q$

Recognisable pattern around the charges (not just in between) *[1 mark]*, lines equally spaced around the charges and joined to the charges, and general symmetry of the diagram *[1 mark]*, arrows along field lines between the charges with arrows pointing away from the positive and towards the negative charge *[1 mark]*.

2) $E = \dfrac{Q}{4\pi\varepsilon_0 r^2}$

$= \dfrac{1.6 \times 10^{-19}}{4\pi \times 8.85 \times 10^{-12} \times (1.75 \times 10^{-10})^2}$

$= 4.698 \times 10^{10}$ *[1 mark]* Vm^{-1} or NC^{-1} *[1 mark]*

3) a) $E = V/d = 1500/(4.5 \times 10^{-3}) = 3.3 \times 10^5$ *[1 mark]* Vm^{-1} *[1 mark]*
   The field is perpendicular to the plates. *[1 mark]*

   b) $d = 2 \times (4.5 \times 10^{-3}) = 9.0 \times 10^{-3}$ m *[1 mark]*
   $E = V/d \Rightarrow V = Ed = [1500/(4.5 \times 10^{-3})] \times 9 \times 10^{-3} = 3000$ V *[1 mark]*

Page 55 — Millikan's Oil-Drop Experiment

1) a) The forces acting on the drop are its weight, acting downwards *[1 mark]* and the equally sized force due to the electric field, acting upwards *[1 mark]*.

   b) Weight = electric force, so $mg = \dfrac{QV}{d}$, and $Q = \dfrac{mgd}{V}$ *[1 mark]*.

   $Q = \dfrac{1.63 \times 10^{-14} \times 9.81 \times 3.00 \times 10^{-2}}{5000} = 9.59 \times 10^{-19}$ C *[1 mark]*

   Divide by the electron charge: $9.59 \times 10^{-19} \div 1.6 \times 10^{-19} = 6$ $\Rightarrow Q = 6e$ *[1 mark]*

   c) The forces on the oil drop as it falls are its weight and the viscous force from the air *[1 mark]*. As the oil drop accelerates, the viscous force increases until it equals the oil drop's weight *[1 mark]*. At this point, there is no resultant force on the oil drop, so it stops accelerating, but continues to fall at terminal velocity *[1 mark]*.

   d) At terminal velocity, $mg = 6\pi\eta rv$. Rearranging, $v = \dfrac{mg}{6\pi\eta r}$ *[1 mark]*
   Find the radius of the oil drop, using mass = volume × density:
   $m = \dfrac{4}{3}\pi r^3 \rho$. So, $r^3 = \dfrac{3m}{4\pi\rho} = \dfrac{3 \times 1.63 \times 10^{-14}}{4 \times \pi \times 880} = 4.42 \times 10^{-18}$
   and $r = 1.64 \times 10^{-6}$ m *[1 mark]*.
   So, $v = \dfrac{1.63 \times 10^{-14} \times 9.81}{6 \times \pi \times 1.84 \times 10^{-5} \times 1.64 \times 10^{-6}} = 2.81 \times 10^{-4} ms^{-1}$ *[1 mark]*

Page 57 — Charged Particles in Magnetic Fields

1) a) $F = Bqv = 0.77 \times 1.6 \times 10^{-19} \times 5 \times 10^6$ *[1 mark]*
   $= 6.16 \times 10^{-13}$ N *[1 mark]*

   b) The force acting on the electron is always at right angles to its velocity, and the speed of the electron is constant. This is the condition for circular motion. *[1 mark]*

2) a) The electron is moving in a circular path, so electromagnetic force = centripetal force
   $Bqv = mv^2 / r$ *[1 mark]*
   Cancelling v from both sides gives: $Bq = mv / r$ *[1 mark]*

   b) $r = \dfrac{mv}{Bq} = \dfrac{9.11 \times 10^{-31} \times 2.3 \times 10^7}{0.6 \times 10^{-3} \times 1.6 \times 10^{-19}} = 0.218$ m *[1 mark]*

Unit 5: Section 3 — Probing Deep into Matter
Page 59 — Scattering to Determine Structure

1) a) The majority of alpha particles are not scattered because the nucleus is a very small part of the whole atom and so the probability of an alpha particle getting near it is small *[1 mark]*. Most alpha particles pass undeflected through the empty space around the nucleus *[1 mark]*.

   b) Alpha particles and atomic nuclei are both positively charged *[1 mark]*. If an alpha particle travels close to a nucleus, there will be a significant electrostatic force of repulsion between them *[1 mark]*. This force deflects the alpha particle from its original path. *[1 mark]*

2) Initial particle energy = 4 MeV
   $= 4 \times 10^6 \times 1.6 \times 10^{-19} = 6.4 \times 10^{-13}$ J *[1 mark]*

   Electrical potential energy $= E_{elec} = \dfrac{Q_{gold}q_{proton}}{4\pi\varepsilon_0 r}$

   $= 6.4 \times 10^{-13}$ J at closest approach. *[1 mark]*

   Rearrange to get $r = \dfrac{(+79e)(+e)}{4\pi\varepsilon_0(6.4 \times 10^{-13})}$ *[1 mark]*

   $= \dfrac{79 \times (1.6 \times 10^{-19})^2}{4\pi \times 8.9 \times 10^{-12} \times 6.4 \times 10^{-13}}$

   $= 2.83 \times 10^{-14}$ m $= 2.8 \times 10^{-14}$ m (to 2 s.f.) *[1 mark]*

Page 61 — Classification of Particles

1) a) X is an electron antineutrino *[1 mark]*.

   b) The electron and the electron antineutrino are leptons *[1 mark]*. Leptons are not affected by the strong interaction, so the decay can't be due to the strong interaction *[1 mark]*.

Answers

Page 65 — Quarks

1)   **udd** [1 mark]
2)   proton = **uud** [1 mark]
     Charge of down quark = –1/3 unit. Charge of up quark = 2/3 unit.
     Total charge = 2/3 + 2/3 – 1/3 = +1 unit [1 mark]

Page 67 — Particle Accelerators

1) $E_{rest} = m_p c^2 = 1.7 \times 10^{-27} \times (3.0 \times 10^8)^2 = 1.53 \times 10^{-10}$ J [1 mark]
     $1.53 \times 10^{-10} \div 1.6 \times 10^{-19} = 9.6 \times 10^8$ eV [1 mark]
     $E_{tot} = 500 \times 10^9$ eV
     So $\gamma = E_{tot} \div E_{rest}$ [1 mark]
          $= 500 \times 10^9 \div 9.6 \times 10^8 \approx 520 \approx 500$ [1 mark]

Page 69 — Electron Energy Levels

1) a) $E = hf = 6.6 \times 10^{-34} \times 4.57 \times 10^{14}$ [1 mark]
        $= 3.0 \times 10^{-19}$ J [1 mark]

   b)
LEVEL	ENERGY	
n = ∞	zero energy	
n = 5	-8.6×10^{-20} J	
n = 4	-1.4×10^{-19} J	
n = 3	-2.4×10^{-19} J	
n = 2	-5.4×10^{-19} J	
n = 1	-2.2×10^{-18} J	[1 mark]

   The difference between these energy levels is 3.0×10^{-19} J, so the electron
   must have fallen between these energy levels.

2 a) 3.8×10^{-5} eV $= 3.8 \times 10^{-5} \times 1.6 \times 10^{-19} = 6.1 \times 10^{-24}$ J
       $\Delta E = hf \Rightarrow f = \Delta E/h = 6.1 \times 10^{-24}/ 6.6 \times 10^{-34}$ [1 mark]
       $= 9.2 \times 10^9$ Hz [1 mark]
   b) 9.2×10^9 oscillations occur every second [1 mark].

Unit 5: Section 4 — Ionising Radiation and Risk

Page 71 — Radioactive Emissions

1) a)

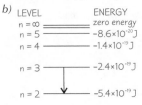

| Skin or paper stops ALPHA | Few mm aluminium stops BETA | Many cm lead stops GAMMA |

   [1 mark for each material stopping correct radiation, total 3 marks]
   b) For 0.6 Gy of alpha, the dose equivalent is 0.6 × 20 = 12 Sv [1 mark].
       For 9 Gy of beta, the dose equivalent is 9 × 1 = 9 Sv [1 mark].
       So, exposure to 0.6 Gy of alpha radiation would be more harmful
       than 9 Gy of beta radiation [1 mark].

Page 73 — Nuclear Decay

1)   The charges on both sides of the equation are balanced [1 mark].

2) a) $^{226}_{88}Ra \rightarrow ^{222}_{86}Rn + ^{4}_{2}\alpha$ [3 marks available — 1 mark for alpha particle,
        1 mark each for proton and nucleon number of radon]
   b) Mass defect = $(6.695 \times 10^{-27}) – (6.645 \times 10^{-27}) = –5.0 \times 10^{-29}$ kg
       [1 mark]. Using the equation $E = mc^2$ [1 mark],
       $E = (–5.0 \times 10^{-29}) \times (3 \times 10^8)^2 = –4.5 \times 10^{-12}$ J
       Energy released = 4.5×10^{-12} J [1 mark]

Page 75 — Binding Energy

1) a) There are 6 protons and 8 neutrons, so the mass of individual parts
        = $(6 \times 1.007276) + (8 \times 1.008665) = 14.112976$ u [1 mark]
        Mass of $^{14}_{6}C$ nucleus = 13.999948 u
        so, mass defect = 13.999948 – 14.112976 = –0.113028 u [1 mark]
        Converting this into kg gives mass defect
        = $–0.113028 \times 1.66 \times 10^{-27}$
        = $–1.876 \times 10^{-28}$ kg $= –1.88 \times 10^{-28}$ kg (to 3 s.f.) [1 mark]
   b) $E = mc^2 = (–1.88 \times 10^{-28}) \times (3 \times 10^8)^2 = –1.69 \times 10^{-11}$ J   [1 mark]
       1 MeV = 1.6×10^{-13} J, so, energy $= \dfrac{-1.69 \times 10^{-11}}{1.6 \times 10^{-13}} = –106$ MeV [1 mark]

2) a) Fusion [1 mark]
   b) The change in binding energy per nucleon is about 0.86 MeV
       [1 mark].
       There are 2 nucleons in ^{2}H, so the change in binding energy is about
       1.72 MeV — so about 1.7 MeV is released (ignoring the positron)
       [1 mark].

Page 77 — Nuclear Fission and Fusion

1) a) Nuclear fission can be induced by neutrons and produces more
        neutrons during the process [1 mark]. This means that each fission
        reaction induces more fission reactions to occur, resulting in an
        ongoing chain of reactions [1 mark].
   b) For example, control rods limit the rate of fission by absorbing
        neutrons [1 mark]. The number of neutrons absorbed by the rods
        is controlled by varying the amount they are inserted into the reactor
        [1 mark]. A suitable material for the control rods in boron [1 mark].
   c) In an emergency shut-down, the control rods are released into the
        reactor [1 mark]. The control rods absorb the neutrons, and stop the
        reaction as quickly as possible [1 mark].

2)   Advantages: keeping the reaction in the nuclear reactor going doesn't
     produce any waste gases that could be harmful to the environment,
     e.g. sulfur dioxide (leading to acid rain) or carbon dioxide [1 mark]. It
     can be used to generate a continuous supply of electricity, unlike
     some renewable sources [1 mark].
     Disadvantages (two of e.g.): problems with the reactor getting out of
     control, risks of radiation from radioactive waste, the emissions
     released in the case of an accident, the long half-life of nuclear waste.
     [1 mark for each disadvantage, maximum 2 marks]

Index

Index